Theory of Periodic Conjugate Heat Transfer

Yuri B. Zudin

Theory of Periodic Conjugate Heat Transfer

With 41 Figures and 11 Tables

 Springer

Professor Dr.-Ing. habil. Yuri B. Zudin
Universitetsky Prospekt
6 (Korpus 4), 32
119333 Moscow, Russian Federation
E-mail: yzudin@gmail.com

Library of Congress Control Number: 2 0 0 7 9 2 8 8 5 5

ISBN 978-3-540-70723-3 Springer Berlin Heidelberg New York

This work is subject to copyright. All rights are reserved, whether the whole or part of the material is concerned, specifically the rights of translation, reprinting, reuse of illustrations, recitation, broadcasting, reproduction on microfilm or in any other way, and storage in data banks. Duplication of this publication or parts thereof is permitted only under the provisions of the German Copyright Law of September 9, 1965, in its current version, and permission for use must always be obtained from Springer. Violations are liable to prosecution under the German Copyright Law.

Springer is a part of Springer Science+Business Media.

springer.com

© Springer-Verlag Berlin Heidelberg 2007

The use of general descriptive names, registered names, trademarks, etc. in this publication does not imply, even in the absence of a specific statement, that such names are exempt from the relevant protective laws and regulations and therefore free for general use.

Typesetting: Data prepared by the Author and SPi
Cover concept and Design: eStudio Calamar Steinen

Printed on acid-free paper   SPIN 11428657   57/3180/SPI   5 4 3 2 1 0

For Tatiana, my wife

Preface

The book was conceived to give the exhaustive answer to a question how thermophysical and geometrical parameters of a body affect heat transfer characteristics under conditions of thermohydrodynamic fluctuations. An applied objective of the book consisted in the development of a universal method of prediction of the average heat transfer coefficient for periodic conjugated processes of heat transfer.

Real "stationary" processes of heat transfer, as a rule, can be considered stationary only on average. Actually, except for a of purely laminar flows, periodic, quasi-periodic and various random fluctuations of parameters (velocities, pressure, temperatures, momentum and energy fluxes, vapor content, interphase boundaries, etc.) about their average values always exist in any fluid flow. Owing to the conjugate nature of the interface "fluid flow–streamlined body," both fluctuation and average values of temperatures and heat fluxes on which surface of heat transfer takes place generally depend on thermophysical and geometrical characteristics of the heat transferring wall. In this connection, a principle question arises about the possible influence of the material and thickness of a wall on the heat transfer coefficient, which is actually the key parameter of convective heat transfer. The facts of such an influence were earlier noticed in experimental investigations of heat transfer at nucleate boiling, dropwise condensation, as well as in some other cases. In these studies, heat transfer coefficients determined as a ratio of the average heat flux on the surface and the average temperature difference between the wall and the fluid could differ noticeably for various materials of the wall (and also for its different thicknesses). These experimental information resulted in the necessity to introduce various correction factors for to "stationary" heat transfer coefficients determined based on the theory of convective heat transfer. As it is believed, such corrections (frequently purely empirical), with any of them being valid for each particular case, brought additional uncertainty in the concept of the average heat transfer coefficient.

Clarity to this question was for the first time brought by my highly respected scientific supervisor Prof. D.A. Labuntsov (1929–1992), who developed a concept of a *true heat transfer coefficient* in 1976. According to this concept, actual values of the heat transfer coefficient (for each point of the heat transferring surface and at each moment of time) are determined solely by hydrodynamic characteristics of the fluid flow and consequently *do not depend* on parameters of the body. Fluctuations of parameters occurring in the fluid flow will result in respective *fluctuations of the true heat transfer coefficient*, also *independent* of the material and thickness of a wall. Then, from a solution of the heat conduction equation with a boundary condition of the third kind, it is possible to find a temperature field in the body (and, hence, on the heat transfer surface) and, as a result, to determine the required *experimental heat transfer coefficient* as a ratio of an average heat flux by an average temperature difference. This value (determined in traditional heat transfer experiments and used in applied calculations) according to its definition should depend on the conjugation parameters.

A study of interrelations of the heat transfer coefficients averaged based on different procedures (*true* and *experimental*) was the main subject of the book of Labuntsov and Zudin (1984) "Processes of heat transfer with periodic intensity," Energoatomizdat, Moscow (in Russian). One of the fundamental results obtained in this book was that the *average experimental value* of the heat transfer coefficient is always less than the *average true value* of this parameter. The present book has arisen as a natural continuation and development of the concept of *true heat transfer coefficient*. The link with the book of Labuntsov and Zudin (1984) is reflected in the first three chapters. The other chapters generalize new results obtained and published by myself since the year of 1991.

Chapter 1 presents a qualitative description of the method proposed as a tool for investigations of periodic conjugate convective–conductive problems "fluid flow–streamlined body." Several particular cases of physical problems including heat transfer processes with periodic fluctuations are outlined briefly in this chapter.

In Chap. 2, an analysis is carried out to a boundary problem for the two-dimensional unstationary heat conduction equation with a periodic boundary condition of the third kind. To characterize quantitatively the thermal effects of a solid body on the average heat transfer, a concept of a *factor of conjugation* is introduced. It is shown that the quantitative effect of the conjugation in the problem can be rather significant.

Chapter 3 outlines a general solution of a boundary problem for the equation of heat conduction with a periodic boundary condition of the third kind. Analytical solutions presented here comprise cases of three characteristic laws of the variation of the true heat transfer coefficient, namely, harmonic, inverse harmonic, and stepwise.

In Chap. 4, a universal algorithm of a general approximate solution of the problem is developed. On its basis, solutions are obtained of several problems for various laws of periodic fluctuations of the true heat transfer coefficient.

Chapter 5 considers conjugate periodic heat transfer for "complex" cases of an external heat supply (or removal): Heat transfer at a contact either with environment or with a second body. A generalized solution for the factor of conjugation for bodies of the "standard form" is obtained. A problem of the conjugate heat transfer for a case of bilateral periodic heat transfer is also investigated in this chapter.

In Chap. 6, an analysis is given for the cases of asymmetric and nonperiodic fluctuations of the true heat transfer coefficient.

Chapter 7 represents some applied problems of the periodic conjugate heat transfer theory such as jet impingement onto a surface, dropwise condensation and nucleate boiling.

The author is deeply grateful to Prof. Wilfried Roetzel (Helmut-Schmidt-Universität, Universität der Bundeswehr Hamburg), and also Prof. Karl Stephan and Prof. Bernhard Weigand (Universität Stuttgart) for the fruitful discussions and their valuable advices promoting an improvement of the overall description of the material outlined in this book.

The preparation of this book would be impossible without the long-term support from the German Academic Exchange Service (DAAD) that repeatedly awarded the author with research grants to undertake research visits to several German universities (Technische Universität München, Universität Paderborn and Universität Stuttgart). The author is very grateful for the financial support by DAAD.

I am also extremely grateful for the continuous and long lasting support of Mrs. Helga Ross. She always believed on the success of my project to write this book and has undertaken a lot of efforts to support me over the last years.

Moscow, July, 2007   Y. B. Zudin

Contents

Abbreviations			XV
Symbols			XVII
1	**Introduction**		1
	1.1	Heat Transfer Processes Containing Periodic Oscillations	1
		1.1.1 Oscillation Internal Structure of Convective Heat Transfer Processes	1
		1.1.2 Problem of Correct Averaging the Heat Transfer Coefficients	3
	1.2	Physical Examples	6
	1.3	Numerical Modeling of Conjugate Convective–Conductive Heat Transfer	10
	1.4	Mechanism of Hydrodynamic Oscillations in a Medium Flowing Over a Body	12
		1.4.1 Van Driest Model	12
		1.4.2 Periodic Model of the Reynolds Analogy	13
		1.4.3 Model of Periodical Contacts	15
	1.5	Hydrodynamic HTC	18
	1.6	Previous Investigations of Heat Transfer Processes with Periodic Intensity	20
	1.7	Analytical Methods	20
	References		21
2	**Construction of a General Solution of the Problem**		27
	2.1	Boundary Value Problem for the Heat Conduction Equation	27
	2.2	Spatial and Temporal Types of Oscillations	30
	2.3	Interrelation between the Two Averaged Coefficients of Heat Transfer	31
	2.4	Dimensionless Parameters	34

XII   Contents

	2.5 Factor of Conjugation: An Analysis of Limiting Variants....	35
	References	36

3   Solution of Characteristic Problems 37
   3.1 Construction of the General Solution..................... 37
   3.2 Harmonic Law of Oscillations 39
   3.3 Inverse Harmonic Law of Oscillations 43
   3.4 Delta-Like Law of Oscillations.......................... 53
   3.5 Step Law of Oscillations................................ 55
   3.6 Comparative Analysis of the Conjugation Effects
        (Smooth and Step Oscillations) 68
   3.7 Particular Exact Solution............................... 69
   References 70

**4   Universal Algorithm of Computation of the Factor
     of Conjugation** 73
   4.1 Smooth Oscillations (Approximate Solutions) 73
   4.2 BC on a Heat Transfer Surface (Series Expansion
        in a Small Parameter)................................... 75
   4.3 Derivation of a Computational Algorithm................. 77
   4.4 Phase Shift Between Oscillations 80
   4.5 Method of a Small Parameter 83
   4.6 Application of the Algorithm for an Arbitrary Law
        of Oscillations... 85
   4.7 Filtration Property of the Computational Algorithm 91
   4.8 Generalized Parameter of the Thermal Effect 92
   4.9 Advantages of the Computational Algorithm 93
   References 93

5   Solution of Special Problems 95
   5.1 Complex Case of Heating or Cooling 95
   5.2 Heat Transfer on the Surface of a Cylinder.............. 102
   5.3 Heat Transfer on the Surface of a Sphere 103
   5.4 Parameter of Thermal Effect for Different Geometrical
        Bodies 104
   5.5 Overall ATHTC 105
        5.5.1 Overall EHTC 105
        5.5.2 Bilateral Spatiotemporal Periodicity of Heat Transfer
              (A Qualitative Analysis) 108
   References 110

**6   Step and Nonperiodic Oscillations of the Heat Transfer
     Intensity** 111
   6.1 Asymmetric Step Oscillations 111
   6.2 Nonperiodic Oscillations................................ 117
   References 120

7	**Practical Applications of the Theory**		121
	7.1	Model Experiment	121
	7.2	Dropwise Condensation	122
	7.3	Nucleate Boiling	126
		7.3.1 Theory of Labuntsov	126
		7.3.2 Periodic Model of Nucleate Boiling	129
	References		136
A	**Proof of the Fundamental Inequalities**		139
	A.1	Proof of the First Fundamental Inequality	139
	A.2	Proof of the Second Fundamental Inequality	145
B	**Functions of the Wall Thickness**		147
	B.1	Spatial Type of Oscillations	148
	B.2	Temporal Type of Oscillations	148
C	**Infinite Chain Fractions**		151
	C.1	Fundamental Theorems of Khinchin	151
	C.2	Generalization of the Third Theorem of Khinchin	152
D	**Proof of Divergence of the Infinite Series**		155
	D.1	Spatial Type of Oscillations	155
	D.2	Temporal Type of Oscillations	156
E	**Functions of Thickness for Special Problems**		159
	E.1	Heat Transfer from the Ambience	159
	E.2	Heat Transfer from an External Semi-Infinite Body	160
Index			161

Abbreviations

ATHTC	Averaged true heat transfer coefficient
BC	Boundary condition
EHTC	Experimental heat transfer coefficient
FC	Factor of conjugation
HTC	Heat transfer coefficient
PTE	Parameter of the thermal effect
TBC	Thermal boundary conditions
THTC	True heat transfer coefficient

Symbols

A_k, A_k^*	Complex conjugate eigenvalues
B_k, B_k^*	Complex conjugate eigenfunctions
b	Amplitude of oscillations of the true heat transfer coefficient
$C_f/2$	Friction factor
c	Specific heat (J kg^{-1} K^{-1})
d_0	Nozzle diameter (m)
F_k	Real parts of eigenfunctions
h	True heat transfer coefficient (THTC) (Wm^{-2} K^{-1})
$\langle h \rangle$	Averaged true heat transfer coefficient (ATHTC) (Wm^{-2} K^{-1})
$\langle \bar{h} \rangle$	Dimensionless averaged true heat transfer coefficient or Biot number
h_m	Experimental heat transfer coefficient (EHTC) (Wm^{-2} K^{-1})
h_0	Steady-state heat transfer coefficient (Wm^{-2} K^{-1})
$\bar{h}_0$	Dimensionless stationary heat transfer coefficient
h_{fg}	Specific enthalpy of evaporation (J kg^{-1})
I_n	Imaginary parts of eigenfunctions
Ja	Jacob number
k	Thermal conductivity (Wm^{-1} K^{-1})
L	Distance between nucleate boiling sites (m)
m	Inverse Fourier number
n_F	Number of boiling sites (m^{-2})
K	Ratio of thermal potentials of contacting media
p	Pressure (Pa)
Pr	Prandtl number
q	Heat flux density (W m^{-2})
$\langle q \rangle$	Averaged heat flux density (W m^{-2})
$\hat{q}$	Oscillating heat flux density (W m^{-2})
q_V	Volumetric heat source (W m^{-3})
R_n	Real parts of eigenvalues
R_*	Critical radius of vapor nucleus (m)

St	Stanton number
t	Dimensionless time
T_s	Saturation temperature (K)
u	Velocity (m c^{-1})
u_0	Free stream velocity (m c^{-1})
u_*	Friction velocity (m c^{-1})
U	Overall heat transfer coefficient (Wm^{-2} K^{-1})
$\langle U \rangle$	Averaged true overall heat transfer coefficient (Wm^{-2} K^{-1})
U_m	Experimental overall heat transfer coefficient (Wm^{-2} K^{-1})
$\langle \bar{U} \rangle$	Dimensionless averaged true overall heat transfer coefficient
E	Generalized factor of conjugation
X	Spanwise coordinate (m)
x	Dimensionless spanwise coordinate
Z	Coordinate along the surface of heat transfer (m)
Z_0	Spatial periods of oscillation (m)
z	Dimensionless coordinate along the heat transfer surface

Greek Letter Symbols

α	Thermal diffusivity (m^2 c^{-1})
Γ	Shear stress (N m^{-2})
δ	Wall thickness (flat plate) (m)
$\bar{\delta}$	Dimensionless wall thickness (flat plate)
δ_f	Thickness of liquid film (m)
ε	Factor of conjugation (FC)
ϑ	Temperature (K)
$\langle \vartheta \rangle$	Averaged temperature (K)
$\hat{\vartheta}$	Oscillating temperature (K)
ϑ_0	Free stream temperature (K)
$\vartheta^\bullet$	Gradient of oscillating temperature (K m^{-1})
ϑ_Σ	Total temperature difference in the three-part system (K)
θ	Dimensionless oscillations temperature
$\theta^\bullet$	Dimensionless gradient of the oscillation temperature (or dimensionless heat flux density)
ξ	Generalized coordinate of a progressive wave
ξ_ϑ	Phase shift between oscillation of true heat transfer coefficient and temperature
ξ_q	Phase shift between oscillation of true heat transfer coefficient and heat flux
μ	Dynamic viscosity (kgm^{-1} s^{-1})
ν	Kinematic viscosity (m^2 s^{-1})
ρ	Density (kg m^{-3})

σ	Surface tension ($\mathrm{N\,m^{-1}}$)
τ	Time (s)
τ_0	Time period of oscillation (s)
Φ_k	Imaginary parts of eigenfunctions
χ	Parameter of thermal effect (PTE)
ψ	Periodic part of the heat transfer coefficient
ω	Frequency ($\mathrm{s^{-1}}$)

Subscripts

+	Active period of heat transfer
−	Passive period of heat transfer
f	Fluid
g	Gas
0	External surface of a body (at $X = 0$)
δ	Heat transfer surface (at $X = \delta$)
min	Minimal value
max	Maximal value
w	Another (second) body

Definition of Nondimensional Numbers and Groups

$\langle \bar{h} \rangle = \frac{\langle h \rangle Z_0}{k}$	Dimensionless averaged true heat transfer coefficient or Biot Number
$\bar{h}_0 = \frac{h_0 Z_0}{k}$	Dimensionless stationary heat transfer coefficient
$E = \frac{U_\mathrm{m}}{\langle U \rangle}$	Generalized factor of conjugation
$Ja = \frac{\rho_\mathrm{f} c_{p\mathrm{f}} \vartheta}{\rho_\mathrm{g} h_\mathrm{fg}}$	Jacob number
$K = \sqrt{\frac{k c \rho}{k_\mathrm{f} c_\mathrm{f} \rho_\mathrm{f}}}$	Ratio of thermal potentials of the contacting media
$m = \frac{Z_0^2}{\alpha \tau_0}$	Inverse Fourier number
$Pr = \frac{\nu_\mathrm{f}}{\alpha_\mathrm{f}}$	Prandtl number
$St = \frac{q}{\rho_\mathrm{f} c_\mathrm{f} u_0 \vartheta_0}$	Stanton number
$\langle \bar{U} \rangle = \frac{\langle U \rangle Z_0}{k}$	Dimensionless averaged true overall heat transfer coefficient
$\bar{U}_\mathrm{m} = \frac{U_\mathrm{m} Z_0}{k}$	Dimensionless experimental overall heat transfer coefficient
$\bar{\delta} = \frac{\delta}{Z_0}$	Dimensionless wall thickness (flat plate)
$\varepsilon = \frac{h_\mathrm{m}}{\langle h \rangle}$	Factor of conjugation

1

Introduction

1.1 Heat Transfer Processes Containing Periodic Oscillations

1.1.1 Oscillation Internal Structure of Convective Heat Transfer Processes

Real stationary processes of heat transfer, as a rule, can be considered stationary only on the average. Actually (except for the purely laminar cases), flows are always subjected to various periodic, quasiperiodic and other casual oscillations of velocities, pressure, temperatures, momentum and energy fluxes, vapor content and interphase boundaries about their average values. Such oscillations can be smooth and periodic (wave flow of a liquid film or vapor, a flow of a fluctuating coolant over a body), sharp and periodic (hydrodynamics and heat transfer at slug flow of a two-phase media in a vertical pipe; nucleate and film boiling process), on can have complex stochastic character (turbulent flows). Oscillations of parameters have in some cases spatial nature, in others they are temporal, and generally one can say that the oscillations have mixed spatiotemporal character.

The theoretical base for studying instantly oscillating and at the same time stationary on the average heat transfer processes are the unsteady differential equations of momentum and energy transfer, which in case of two-phase systems can be notated for each of the phases separately and be supplemented by the conditions of the physical interface on the boundaries between phases (conditions of conjugation). An exhaustive solution of the problem could be a comprehensive analysis with the purpose of a full description of any particular fluid flow and heat transfer pattern with all its detailed characteristics, including various fields of oscillations of its parameters.

However, at the time being such an approach can not be realized in practice. The problem of modeling turbulent flows [1] can serve as a vivid example.

As a rule at its theoretical analysis, Reynolds-averaged Navier–Stokes equations are considered, which describe time-averaged quantities of fluctuating parameters, or in other words turbulent fluxes of the momentum and energy. To provide a closed description of the process, these correlations by means of various semiempirical hypotheses are interrelated with time-averaged fields of velocities and enthalpies. Such schematization results in the statement of a stationary problem with spatially variable coefficients of viscosity and thermal conductivity. Therefore, as boundary conditions here, it is possible to set only respective stationary conditions on the heat transfer surface of such a type as, for example, "constant temperature," "constant heat flux."

It is necessary to specially note, that the replacement of the full "instant" model description with the time-averaged one inevitably results in a loss of information on the oscillations of fluid flow and heat transfer parameters (velocities, temperatures, heat fluxes, pressure, friction) on a boundary surface. Thus the theoretical basis for an analysis of the interrelation between the temperature oscillations in the flowing ambient medium and in the body is omitted from the consideration. And generally saying, the problem of an account for possible influence of thermophysical and geometrical parameters of a body on the heat transfer at such a approach becomes physically senseless. For this reason, such a "laminarized" form of the turbulent flow description is basically not capable of predicting and explaining the wall effects on the heat transfer characteristics, even if these effects are observed in practice. The problem becomes especially complicated at imposing external oscillations on the periodic turbulent structure that takes place, in particular, in flows over aircraft and spacecraft. Unresolved problems of closing the Navier–Stokes equations in combination with difficulties of numerical modeling make a problem of detailed prediction of a temperature field in the flowing fluid very complicated. In some cases, differences between the predicted and measured local *heat transfer coefficient* (HTC) exceeds 100%.

In this connection the direction in the simulation of turbulent flows based on the use of the primary transient equations [2] represents significant interest. The present book represents results of numerical modeling of the turbulent flows in channels subjected to external fields of oscillations (due to vortical generators etc.). It is shown that in this case an essentially anisotropic and three-dimensional flow pattern emerges strongly different from that described by the early theories of turbulence [1]. In the near-wall zone, secondary flows in the form of rotating "vortical streaks" are induced that interact with the main flow. As a result, oscillations of the thermal boundary layer thickness set on, leading to periodic enhancement or deterioration of heat transfer. Strong anisotropy of the fluid flow pattern results in the necessity of a radical revision of the existing theoretical methods of modeling the turbulent flows. So, for example, the turbulent Prandtl number being in early theories of turbulence [1] a constant of the order of unity (or, at the best, an indefinite scalar quantity), becomes a tensor.

It is necessary to emphasize that all the mentioned difficulties are related to the nonconjugated problem when the role of a wall is reduced only to maintenance of a *boundary condition* (BC) on the surface between the flowing fluid and the solid wall.

1.1.2 Problem of Correct Averaging the Heat Transfer Coefficients

The basic applied task of the book is the investigation into the effects of a body (its thermophysical properties, linear dimensions and geometrical configuration) on the traditional HTC, measured in experiments and used in engineering calculations. Processes of heat transfer are considered stationary on average and fluctuating instantly. A new method of investigation of the conjugate problem "fluid flow–body" is presented. The method is based on a replacement of the complex mechanism of oscillations of parameters in the flowing coolant by a simplified model employing a varying "true heat transfer coefficient" specified on a heat transfer surface.

The essence of the developed method can be explained rather simply. Let us assume that we have perfect devices measuring the instant local values of temperature and heat fluxes at any point of the fluid and heated solid body. Then the hypothetical experiment will allow finding the fields of temperatures and heat fluxes and their oscillations in space and in time, as well as their average values and all other characteristics. In particular, it is possible to present the values of temperatures (exactly saying temperature heads or loads, i.e., the temperatures counted from a preset reference level) and heat fluxes on a heat transfer surface in the following form:

$$\vartheta = \langle \vartheta \rangle + \hat{\vartheta}, \qquad (1.1)$$

$$q = \langle q \rangle + \hat{q}, \qquad (1.2)$$

i.e., to write them as the sum of the averaged values and their temporal oscillations. For the general case of spatiotemporal oscillations of characteristics of the process, the operation of averaging is understood here as a determination of an average with respect to time τ and along the heat transferring surface (with respect to the coordinate Z). The *true heat transfer coefficient* (THTC) is determined on the basis of (1.1) and (1.2) according to Newton's law of heat transfer [3, 4]:

$$h = \frac{q}{\vartheta}. \qquad (1.3)$$

This parameter can always be presented as a sum of an averaged part and a fluctuating additive:

$$h = \langle h \rangle + \hat{h}. \qquad (1.4)$$

It follows from here that the correct averaging of the HTC is as follows

$$\langle h \rangle = \left\langle \frac{q}{\vartheta} \right\rangle. \tag{1.5}$$

Therefore we shall call parameter $\langle h \rangle$ an *averaged true heat transfer coefficient* (ATHTC). The problem consists in the fact that the parameter $\langle h \rangle$ cannot be directly used for applied calculations, since it contains initially the unknown information of oscillations $\hat{\vartheta}, \hat{q}$. This fact becomes evident if (1.5) is rewritten with the help of (1.1) and (1.2):

$$\langle h \rangle = \left\langle \frac{\langle q \rangle + \hat{q}}{\langle \vartheta \rangle + \hat{\vartheta}} \right\rangle. \tag{1.6}$$

The purpose of the heat transfer experiment is the measurement of averaged values of an averaged temperature $\langle \vartheta \rangle$ and a heat flux $\langle q \rangle$ on the surfaces of a body and determination of the traditional HTC

$$h_{\mathrm{m}} = \frac{\langle q \rangle}{\langle \vartheta \rangle}. \tag{1.7}$$

The parameter h_{m} is fundamental for carrying out engineering calculations, designing heat transfer equipment, composing thermal balances, etc. However it is necessary to point out that transition from the initial Newton's law of heat transfer (1.3) to the restricted (1.7) results in the loss of the information of the oscillations of the temperature $\hat{\vartheta}$ and the heat fluxes $\hat{q}$ on the wall.

Thus, it is logical to assume that the influence of the material and the wall thickness of the body taking part in the heat transfer process on HTC h_{m} uncovered in experiments is caused by noninvariance of the value of h_{m} with respect to the Newton's law of heat transfer. For this reason we shall refer further to the parameter h_{m} as to an *experimental heat transfer coefficient* (EHTC).

Thus, we have two alternative procedures of averaging the HTC: true (1.5) and experimental (1.7). The physical reason of the distinction between $\langle h \rangle$ and the h_{m} can be clarified with the help of the following considerations:

- Local values $\langle \vartheta \rangle$ and $\langle q \rangle$ on a surface where heat transfer takes place are formed as a result of the thermal contact of the flowing fluid and the body.
- Under conditions of oscillations of the characteristics of the coolant, temperature oscillations will penetrate inside the body.
- Owing to the conjugate nature of the heat transfer in the considered system, both fluctuating $\hat{\vartheta}, \hat{q}$, and averaged $\langle \vartheta \rangle, \langle q \rangle$ parameters on the heat transfer surface depend on the thermophysical and geometrical characteristics of the body.
- The ATHTC $\langle h \rangle$ directly follows from Newton's law of heat transfer (1.3) (which is valid also for the unsteady processes) and consequently it is determined by hydrodynamic conditions in the fluid flowing over the body.

- The EHTC h_m by definition does not contain the information on oscillations $\hat{\vartheta}, \hat{q}$, and consequently it is in the general case a function of parameters of the interface between fluid and solid wall.
- Aprioristic denying of dependence of the EHTC on material properties and wall thickness is wrong, though under certain conditions quantitative effects of this influence might be insignificant.

From the formal point of view, the aforementioned differences between the true (1.5) and experimental (1.7) laws of averaging of the actual HTC is reduced to a rearrangement of the procedures of division and averaging. This situation is illustrated evidently in Fig. 1.1.

Using the concepts introduced above, the essence of a suggested method can be explained rather simply. We shall assume that for the case under investigation the HTC h is known: $h = h(Z, \tau)$, where Z and τ are the coordinate along a surface where heat transfer takes place and the time, respectively. According to the internal structure of the considered processes this parameter should have periodic, quasiperiodic, or generally fluctuating nature, varying

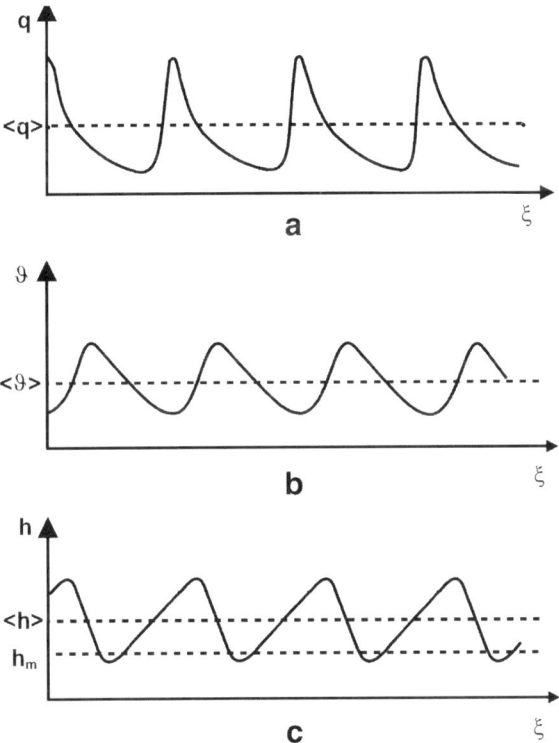

Fig. 1.1. True and experimental laws of the averaging of the heat transfer coefficient; (**a**) heat flux density on the heat transfer surface, (**b**) temperature difference "wall–ambience," and (**c**) heat transfer coefficient

about its average value $\langle h \rangle$: $h = \langle h \rangle + \hat{h}(Z,\tau)$. This information is basically sufficient for the definition of actual driving temperature difference $\vartheta(Z,\tau)$ heat fluxes $q(Z,\tau)$ in a massive of a heat transferring body, and, hence, on the heat transfer surface. Thus, the calculation is reduced to a solution of a boundary value problem of the unsteady heat conduction equation [5]

$$\frac{\partial \vartheta}{\partial \tau} = \alpha \left(\frac{\partial^2 \vartheta}{\partial X^2} + \frac{\partial^2 \vartheta}{\partial Z^2} \right) + \frac{q_V}{c\rho}, \qquad (1.8)$$

with the boundary condition (BC) of the third kind on the heat transfer surface

$$-k \frac{\partial \vartheta}{\partial X} = h\vartheta \qquad (1.9)$$

and suitable BC on the external surfaces of the body.

It is essential for our analysis that up to the same extent in which the information about the function $h = h(Z,\tau)$ is trustworthy, the computed parameters $\vartheta(Z,\tau)$ and $q(Z,\tau)$ are determined also authentically. The basis for such a statement is the fundamental theorem of uniqueness of the solution of a boundary value problem for the heat conduction equation [5]. In other words, the temperature field ϑ and heat fluxes q found in the calculation should appear identical to the actual parameters, which could be in principle measured in a hypothetical experiment. Further based on the known distributions ϑ and q, it is possible to determine corresponding average values $\langle \vartheta \rangle$ and $\langle q \rangle$, and finally (from (1.7)) the parameter h_m, which appears to be a function of the parameters of conjugation. It follows from the basic distinction of procedures of averaging of (1.5) and (1.7) that an experimental value of the actual HTC is not equal to its averaged true value:

$$h_m \neq \langle h \rangle, \qquad (1.10)$$

The analytical method schematically stated above, in which "from the hydrodynamic reasons" the following relation is stated

$$h(z,\tau) = \langle h \rangle + \hat{h}(Z,\tau), \qquad (1.11)$$

and further from the solution of the heat conduction equation in a body the parameter h_m is determined, outlines the basic essence of the approach developed in the present book. Different aspects of this method are discussed below in more detail.

1.2 Physical Examples

For the practical realization of this method it is necessary for each investigated process to specify the parameter $h(Z,\tau)$ (i.e., THTC) periodically varying with respect to its average value. A difficulty thus consists in the fact

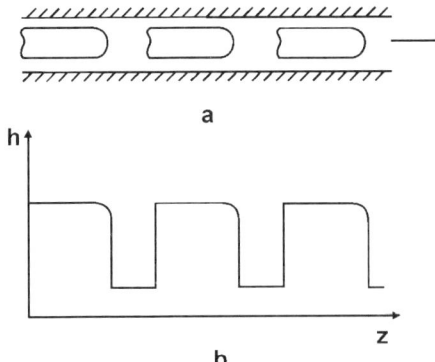

Fig. 1.2. Slug flow of a two-phase fluid: (**a**) schematic of the process, (**b**) variation of the THTC with the longitudinal coordinate

that, generally speaking, a valid function outlining the change of the THTC (with all its details) is unknown for any real periodic process. Therefore, the specification of this parameter is possible only approximately. This freedom in choice of the THTC inevitably makes results of the analysis dependent on the accepted approximations and assumptions. Thus the approximate nature of the developed method consists namely in this aspect. From the mathematical point of view, all constructions, solutions, estimations, and conclusions are obtained quite strictly and precisely. Physical features of some characteristic processes of heat transfer with periodic oscillations are discussed below.

Slug flow of a two-phase medium. A schematic image of this type of flow frequently met in engineering applications is given in Fig. 1.2. Oscillations of the heat transfer intensity in each section of the channel are caused here by the periodic passage of a large steam bubble and a liquid volume. Instant picture of the HTC variation over the height of a pipe is shown in the same figure. The thickness of the liquid film δ_f formed on a wall during passage of a steam bubble, can be determined using known recommendations documented in [6, 7]. The THTC is practically equal to thermal conductivity of a liquid layer k_f/δ_f, where k_f is the heat conductivity of the liquid phase. During the passage of the liquid, the heat transfer intensity is determined by the relations for heat transfer to a turbulent flow. Thus the character of the variation of the THTC with respect to time and to the vertical coordinate can be considered periodic step function. The curve of $\delta_f(Z,\tau)$ here will move upward with speed of movement of the steam bubbles along the wall of a pipe. For the considered case, it is essential that the function $h(Z,\tau)$ is determined by fluid flow peculiarities in the two-phase medium and consequently does not depend on the thermophysical properties and thickness of the wall.

Flow over a body in the vicinity of the stagnation point. The schematization of this type of flow is shown in Fig. 1.3. It is easy to show that at presence of the periodic oscillations of the velocity of a fluid about its average value, the heat

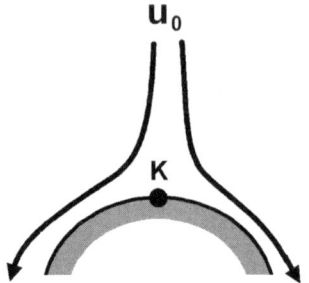

Fig. 1.3. Flow over a body in the vicinity of a stagnation point

transfer intensity will be also periodic in time. In other words, if the period of change in the fluid velocity is essentially larger than the time needed for the individual particles of a liquid to pass by zone where heat transfer is studied (in the vicinity of the frontal stagnation point K), the instant behavior of heat transfer can be considered quasistationary, with the function $h(\tau)$ being equal to the stationary dependence $h[u_0(\tau)]$.

In the considered case, the time variation of the heat transfer intensity follows from the hydrodynamic conditions of flow, and THTC remains actually constant for various materials of the surface.

Flow in a laminar boundary layer. Let us consider stationary flow in a laminar boundary layer on which periodic velocity oscillations are imposed. From the same reasons, as in the example of the fluid flow over a body in the vicinity of a stagnation point considered above, the process of heat transfer here can be considered quasistationary: $h(\tau) = h[u_0(\tau)]$. For a case where the amplitude of the velocity oscillations is comparable to the velocity's average value, it is necessary to expect backward influence of the imposed oscillations on the average level of heat transfer. As known [4], a stationary HTC h_0 in a laminar boundary layer depends on the velocity as

$$h_0 = C\sqrt{u}. \qquad (1.12)$$

Here $C = 0.332\rho_\mathrm{f} c_\mathrm{f}/Pr^{2/3}\sqrt{\nu_\mathrm{f}/X}$, X is the distance from the initial stagnation point of a plate. Imposing of harmonic velocity oscillations on the stationary flow $u \to \langle u \rangle [1 + b\cos(2\pi\tau/\tau_0)]$ results in corresponding oscillations of the THTC $h_0 \to h_0(1 + \tilde{h})$, so that (1.12) takes the following form:

$$h_0(1+\tilde{h}) = C\sqrt{\langle u \rangle [1 + b\cos(2\pi\tau/\tau_0)]}. \qquad (1.13)$$

Averaging (1.13) over the period of oscillations τ_0 gives:

$$h = Cf(b)h_0. \qquad (1.14)$$

Here $f(b)$ is a rather complex function of the oscillations amplitude, which weakly decreases with increasing b: $b = 0, f(b) = 1; b = 1, f(b) \approx 0.9$. Subtracting (1.14) of (1.13) term-by-term, one can find the fluctuating component of the THTC. In the case of negligibly small amplitude $b \to 0$, these oscillations will look like as a cosine function:

$$h_0 = C(b/2)\cos(2\pi\tau/\tau_0). \tag{1.15}$$

In a limiting case of the maximal amplitude $b = 1$, it can be deduced from (1.13):

$$h_0 = C\left[\pi/2\left|\cos(\pi\tau/\tau_0)\right| - 1\right]. \tag{1.16}$$

As it is obvious from (1.16), at transition from $b \to 0$ to $b = 1$ oscillations of the heat transfer intensity are strongly deformed: the period decreases twice, and the form sharpens and is pointed from top to bottom. On the other hand, the average heat transfer level changes thus only by $\cong 10\%$: at maximal amplitude ($b = 1$) the ATHTC equals to $h \approx 0.9h_0$. Thus, the strong change in the amplitude of oscillations leads only to minor change of the average heat transfer level.

Wave flow of a liquid film. At film condensation of a vapor on a vertical surface and also at evaporation of liquid films flowing down, one can observe a wave flow of the film already at small values of the film Reynolds numbers [6,7]. Under these conditions, the wavelength essentially exceeds the film thickness, and the phase speed of its propagation is of the same order as the average velocity of the liquid in the film. As the Reynolds numbers increase, the character of flow changes: a thin film of a liquid of approximately constant thickness is formed on the surface, on which discrete volumes of a liquid periodically roll down. At a wave mode of the film flow, the THTC is rather precisely described by the dependence $h(Z,\tau) = k_f/\delta_f(Z,\tau)$ specified for the first time by Kapitsa in his pioneer works [8,9]. It follows from this dependence that at a harmonic film structure the THTC is characterized by an inverse harmonic function (Fig. 1.4). At a flow with a "rolling down" liquid, a description of the THTC can be constructed similarly to the case of the slug flow of a two-phase medium considered above, i.e., also independently of the

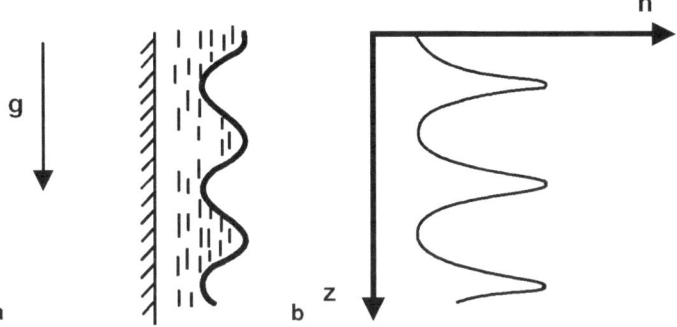

Fig. 1.4. Wave flow of a liquid film: (a) schematic of the process (b) variation of the THTC with the longitudinal coordinate (g is the gravitational acceleration)

thermal influence of a solid body. At a wave mode of condensation of vapor of liquid metals (sodium, potassium), nonequilibrium molecular-kinetic effects in the vapor phase play a significant role, due to the process of capturing (condensation) of the molecules of vapor. Therefore for a calculation of the heat transfer for vapor condensation (as well as for liquid film evaporation) of a liquid metal, these effects should be taken into account together with the thermal resistance of the liquid film itself determined by the formula of Kapitsa.

Near-wall turbulent flows. The structure of the hydrodynamic oscillations in the turbulent flows are very complex and include a wide spectrum of oscillations with various scales and amplitudes. Along with the so-called stochastic noise, typical for casual processes in a flow, there exist also large-scale periodic oscillations caused by periodic entrainment of accelerated portions of a fluid from the core of the flow into the near-wall region. The average time intervals between these periodic entrainments, and also characteristics of oscillations of the wall friction have been determined in a number of experimental investigations (see, e.g., [10, 11]). On the basis of the Reynolds analogy, it is possible to expect that the wall heat flux will undergo also similar oscillations. It is essential for our analysis that oscillations of parameters are connected with the movement of large turbulent vortical streaks and are consequently caused by the hydrodynamics of the flow. It is again obvious in the examined case that the THTC is independent of the material of a solid body.

1.3 Numerical Modeling of Conjugate Convective–Conductive Heat Transfer

The needs of modern engineering applications (in particular, aerospace engineering) dictate extremely strict requirements for termally loaded surfaces and of critical conditions of the flow aerodynamics. In order to meet these requirements, it is necessary to have an effective tool for the solution of various problems of conjugate convective–conductive heat transfer. Numerical modeling of the velocity field in a fluid flow as well as conjugated temperature fields in a solid body and in the fluid was carried out in [12]. For the calculation of temperature fields at any spatial location and at any moment of time, a Finite-Element method was used. Compact representation of the conjugated fields of temperatures as a uniform symmetric matrix has allowed the author of the work [12] to carry out an effective calculation of a solid body and a fluid for different geometries, thermophysical properties and conditions of heat transfer. Thus, the necessary information on the distribution of temperatures along the heat transfer surface for a number of applied problems (a supersonic flow over an aircraft, flow in compact heat exchangers of a complex configuration, a three-dimensional flow around turbine airfoils etc.) can be obtained. The problem of the thermal interface "fluid flow–body" was schematically

1.3 Numerical Modeling of Conjugate Convective–Conductive Heat Transfer

represented in [12] as "an aerodynamic triangle." This triangle shows that in any case an interaction between two components takes place, while the third component remains passive. Possible pair interactions are listed below:

- Ambient medium (fluid) and a body cooperate by means of friction and convection. The fluid determines the quantitative, but not the qualitative character of interaction.
- Interaction of a fluid and a body is determined by combination of their thermophysical properties (for example, viscosity and density), and also the nature of a fluid (liquid, gas, or a two-phase stream) independently of a solid body bordering with the fluid.
- A fluid and a body interact through temperature fields and "catalytic effects" independently of the flow regime (laminar or turbulent, incompressible or compressible, etc.).

Ideally, an analysis on the basis of the aerodynamic triangle is called to give an exhaustive description of any conjugate problem. However, as it is pointed in [12], in practice in a real numerical experiment only separate parts (or "legs") of the triangle are used. In other words, by modeling of the particular conjugate problem one should distinguish the main characteristic feature (turbulence, unsteadiness, chemical reactions, etc.). Depending on this, respective simplifications of the mathematical description will be further carried out: linearization of separate terms, replacement of the numerical solution of the system of equations by iterative procedure etc. Thus, the initially global structure of a numerical method results in practice in the necessity of particularly relevant approximations, estimations, neglecting of terms, etc. An application of the specified approximations within the framework of an apparently strict and self-sufficient numerical method is explained in [12] by the primary approximate nature of the used discrete numerical methods, and also by the necessity of minimization the computational time. These inherent features of numerical methods persist until now, despite of the rapid development of these methods over the last decades. Ideologically rather similar to [12] numerical research of the conjugate problem fluid flow–body has been carried out in [13].

As a conclusion, one can note that by modeling of the conjugate systems fluid flow–body in [12,13] important and interesting results have been obtained allowing, in particular, to analyze temperature fields in different interacting media. However, the authors of [12, 13] have not dealt with the problem of averaging of the actual HTC at presence of periodic oscillations in the flow (as well as they have not addressed the whole range of issues associated with this problem and discussed in the present book). As we believe, the reason for this lies not in the computational (mathematical) aspects of the problem, but in the issues that have fundamental (physical) character. On the one hand, the use of the rapidly developing modern computer codes indeed allow solving effectively two- and three-dimensional unsteady transport equations for the conjugated media. On the other hand, as far as it is known to the author, any comprehensive technique has not been created so far that could allow displaying real oscillations of thermohydraulic parameters as respective terms in the transport equations.

Meanwhile there is an urgent need for the everyday engineering and thermophysical practice in creation of a justified tool for a reliable prediction of the thermal energy transfer at the presence of periodic oscillations of thermohydraulic parameters in the flow. So, for example, the account for the dependence of the heat transfer intensity at nucleate boiling of a liquid on the thermophysical properties of a body till now is carried out on the basis of empirical recommendations of [14]. The listed reasons testify in favor of the benefit of the approximate method of the analysis of the periodic connected heat transfer developed in the present book.

1.4 Mechanism of Hydrodynamic Oscillations in a Medium Flowing Over a Body

1.4.1 Van Driest Model

Let us consider the known model of Van Driest [4] describing the law of attenuation of the velocity oscillations in the near-wall region of a turbulent flow. The model is based on the classical exact solution of the Navier–Stokes equations (second problem of Stokes [15]). Consideration is given to an unsteady multilayer flow caused by harmonic oscillations (with the frequency ω) of an infinite solid surface around its own plane. By the virtue of the no-slip BC on the surface, oscillation of the wall results in the fact that the fluid on the solid surface of interface ($y = 0$) possesses some velocity varying under the law:

$$y = 0 : u(0, \tau) = u_0 \cos(\omega \tau). \tag{1.17}$$

The system of the Navier–Stokes equations is reduced to one equation for the longitudinal velocity, with its convective terms being identically equal to zero:

$$\frac{\partial u}{\partial \tau} = \nu_f \frac{\partial^2 u}{\partial y^2}. \tag{1.18}$$

Solution of (1.18) with the BC (1.17) results in:

$$u(y, \tau) = u_0 \exp\left(-y\sqrt{\frac{\omega}{2\nu_f}}\right) \cos\left(\omega\tau - y\sqrt{\frac{\omega}{2\nu_f}}\right). \tag{1.19}$$

According to (1.19), the fluid performs oscillations with amplitude decreasing away of the wall

$$u = u_0 \exp\left(-y\sqrt{\frac{\omega}{2\nu_f}}\right). \tag{1.20}$$

Oscillations of the fluid layer, which is located at the distance y counted from the wall, have a phase shift $y\sqrt{\omega/(2\nu_f)}$ in comparison to the oscillations at a wall. The phase shift is directed opposite to movement of the wall. As the surface $y = 0$ is actually at rest, a flow corresponding to synchronous oscillations of the whole infinite volume of a fluid with the velocity $u_0 \cos(\omega\tau)$ is imposed on the obtained flow. This means that in order to provide the required character of the velocity oscillations, an indefinitely extended source of momentum is entered into the right-hand side of (1.18) without any substantial justifications. Extension of the Van Driest scheme for the problem of attenuation of the temperature oscillations results in the necessity of introduction of the similar nonphysical source terms in the energy equation for the fluid. At last, an attempt to state the conjugate problem based on the similar approach results in the physically absurd introduction of virtual thermal sources both in the fluid, and in the body.

As far as it is known to the author of the present book, the mentioned obvious incorrectness of the widely known Van Driest model has not been commented anyhow in the literature. It once again confirms the conclusion that a correct statement of the problem of conjugation of temperature fields in the environment and in a body in view of a real behavior of oscillations (as well as the derivation of its solution) encounters serious difficulties. In this connection, correct approximate models of thermohydraulic processes with periodic intensity gain more importance. A simple model describing interrelation of laws of friction and heat transfer in the turbulent near-wall flow is stated below.

1.4.2 Periodic Model of the Reynolds Analogy

As it is known, for flow in a turbulent boundary layer for $Pr = 1$ a similarity of the longitudinal velocity and temperature fields takes place, from which the classical Reynolds analogy [3, 4, 15] follows

$$St = C_f/2. \tag{1.21}$$

Here

$$St = \frac{q}{\rho_f c_f u_0 \vartheta_0} \quad \text{and} \quad C_f/2 = \frac{\Gamma}{\rho_f u_0^2}, \tag{1.22}$$

are the Stanton number and friction coefficient, respectively; q is the heat flux density; Γ is the shear stress. At $Pr \neq 1$, the similarity of the velocity and temperature distributions holds for a turbulent core of the flow, however it is broken in the near-wall region. This case, which is described within the framework of different schemes of the so-called extended Reynolds analogy, results in the use of different correction factors in the right-hand side of (1.21). These corrections are determined, as a rule, with the help of rather labor-consuming procedures (introduction of the radial velocity distributions and friction coefficients, calculation of the Lyon's integrals, etc.). Known correlation for the extended Reynolds analogy [3] looks like

$$St = \frac{C_f/2}{1 + 11.7\sqrt{C_f/2}\left(Pr^{2/3} - 1\right)}. \quad (1.23)$$

Let us show that expressions like (1.23) can be derived from a simple flow model describing the interaction between a wall and a flow periodically entrained from the core of the accelerating cold fluid flow. A physical basis of this model is the phenomenon of the above mentioned "bursting" described in [10,11]. These works mentioned for the first time the existence in near-wall regions of flow of specific coherent structures in form of pair vortices extended in the direction of flow and periodically pushed out into the turbulent core of the fluid. Let us accept that after collisions with a wall the homogeneous volume of a fluid with parameters u_0, ϑ_0 continues moving downstream, leaving on the wall its trace in the form of a laminar boundary layer (Fig. 1.5). Velocity and temperature difference on external boundary of the near-wall layer will be equal to $u_\delta, \vartheta_\delta$, respectively. Let us write down the known laws of friction and heat transfer for a laminar boundary layer [3]

$$\Gamma_\delta(Z) = A(Z)\rho_f u_\delta^2, q_\delta(Z) = \frac{A(Z)}{Pr^{2/3}}\rho_f c_f u_\delta \vartheta_\delta. \quad (1.24)$$

Here $A(Z) = 0.332/\sqrt{Re_Z}$, $Re_Z = u_\delta Z/\nu_f$ is the local Reynolds number. In accordance with the phenomenon of "bursting," after a certain time period τ_0 there should be a replacement of the fluid volume drifting over a wall by the new volume invading into the near-wall layer from the turbulent core flow. During this time period, individual particles of the fluid in the laminar wake of the previous fluid volume reach a certain coordinate $Z_0 = u_\delta \tau_0$. The subsequent emission of the decelerated heated fluid from the near-wall region and its replacement with a new portion of the accelerated cold fluid will lead

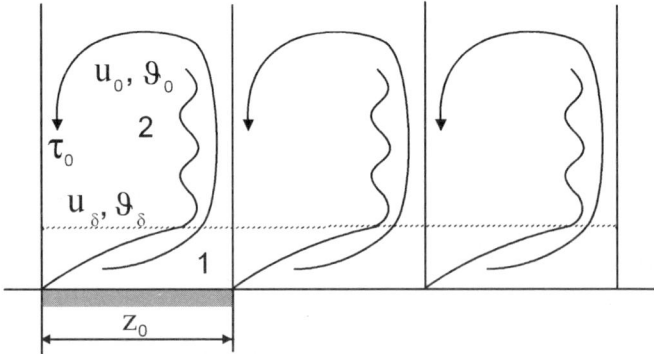

Fig. 1.5. Schematic of the near-wall turbulent flow: 1 – laminar boundary layer, 2 – turbulent core of the flow

1.4 Mechanism of Hydrodynamic Oscillations

to a renewal of a laminar boundary layer on the wall and a repetition of all the subsequent actions. On the external boundary of the near-wall layer, there will be momentum and heat exchange with the fluid invading from the turbulent core. This process can be approximately described with the one-dimensional transient equations for the differences of velocities $u_0 - u_\delta$ and temperatures $\vartheta_0 - \vartheta_\delta$ on border of semi-infinite bodies

$$\Gamma_0(\tau) = \mu_f \frac{u_0 - u_\delta}{\sqrt{\pi \nu_f \tau}}, \quad q_0(\tau) = k_f \frac{\vartheta_0 - \vartheta_\delta}{\sqrt{\pi \alpha_f \tau}}. \tag{1.25}$$

According to the described model, spatial (in near-wall regions) and temporal (in the core of the flow) periodic flow pattern exists. Natural conditions of the interface between these regions will be the equality of the respective time-averaged (with respect to spatial Z_0 and time τ_0 scales) momentum and heat fluxes

$$\langle \Gamma_\delta(Z) \rangle = \langle \Gamma_0(\tau) \rangle = \Gamma, \quad \langle q_\delta(Z) \rangle = \langle q_0(\tau) \rangle = q. \tag{1.26}$$

Then from (1.22) to (1.26) it is possible to obtain a correlation for the extended Reynolds analogy:

$$St = \frac{C_f/2}{\sqrt{Pr}\left[1 + \sqrt{(C_f/2)/\langle A \rangle}\left(Pr^{1/6} - 1\right)\right]}, \tag{1.27}$$

where $\langle A \rangle = 0.664/Re_{Z_0}$, $Re_{Z_0} = u_\delta Z_0/\nu_f$. For the expression (1.27) to pass to (1.23) in the limiting case of $Pr \to \infty$, it is necessary to put: $\langle A \rangle = 1/11.7^2$. It is interesting to note, that at values of $Pr \geq 1$ correlation (1.27) reduces to the relation

$$St \approx \frac{C_f/2}{\sqrt{Pr}}, \tag{1.28}$$

which agrees well with the solution given in [16]. The resulting simple model evidently illustrates the physical expediency of taking into account of internal fluctuating structures in real heat transfer processes.

1.4.3 Model of Periodical Contacts

A simple evident model of the conjugate problem fluid flow–body is a scheme of periodic collisions with a surface of a solid body (conductive supply of heat into the system) of the volumes of fluid constantly replacing each other (convective removal of heat) – Fig. 1.6. Since a constant heat flux is supplied from the depth of a solid body, the distribution of the average temperature in the body should look like linear functions. On this linear distribution, temperature oscillations with increasing amplitude (as approaching to the surface) will be imposed. In doing so, the "conductive condition of periodicity" should be fulfilled: temperature distribution in the solid body at time $\tau = \tau_0$ should exactly repeat the respective distribution at time $\tau = 0$. The temperature

16   1 Introduction

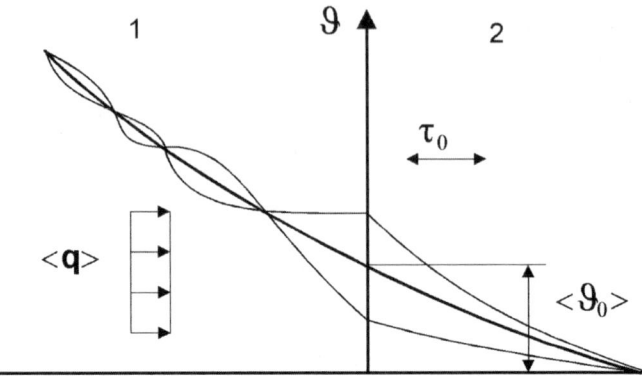

Fig. 1.6. Schematic of the periodical contacts of two media: 1 – body, 2 – ambient fluid

of a surface of the next cold fluid volume will always grow in time (stepwise at the initial moment of time, and then as a monotonic function during the entire period of interaction). The "convective condition of periodicity" will be expressed in the replacement of a heated volume after the end of the interaction of a wall with a new cold volume. The mathematical description of the problem includes the unsteady one-dimensional equations of heat conduction for the solid body and the volume of fluid completed with the conditions of conjugation at the interface (equality of temperatures and heat fluxes). The described model of periodic contacts contains a unique dimensionless parameter, which is the ratio of thermal potentials of the contacting media $K = \sqrt{(kc\rho)/(k_\mathrm{f} c_\mathrm{f} \rho_\mathrm{f})}$. Nevertheless, apparent simplicity of the problem is deceptive. Its solution with the help of the Green's function [5] results in obtaining a complex integro-differential equation. Let us consider the heat conduction equation for a volume of fluid for the limiting cases allowing an analytical solution:

(a) ϑ = const. The limiting case for ϑ = const will be reached for $K \to \infty$. In this case, oscillations of temperatures and averaged temperature gradient in a body will be negligibly small. The known solution [5] for a case ϑ = const gives: $q = k_\mathrm{f} \vartheta / \sqrt{\pi \alpha_\mathrm{f} \tau}$. It follows from here that the heat flux averaged over the period of contact t_0 will be equal to $\langle q \rangle = 2 k_\mathrm{f} \vartheta / \sqrt{\pi \alpha_\mathrm{f} \tau_0}$. Under these conditions, the EHTC and ATHTC will be equal to each other: $h_\mathrm{m} = \langle h \rangle = 2 k_\mathrm{f} / \sqrt{\pi \alpha_\mathrm{f} \tau_0}$. One should note that in the general case final values of the complex K under conditions of conjugation of a flowing fluid and body temperature oscillations will penetrate into the body, and the isothermal wall condition thus will be broken.

(b) q = const. In the other limiting case $K \to 0$, temperature oscillations in a body will reach their maximum. It follows from the Fourier law that at $k \to 0$ an infinitely large average temperature gradient corresponds to a final average heat flux in a body. This means from a physical point of view an unlimited increase in the heat flux rate, in relation to which any finite oscillations will be considered negligibly small. This corresponds to a limiting case $q = $ const. The known solution [5] for this case gives the law of monotonic increase of temperature in time: $\vartheta = 2q\sqrt{\alpha_f \tau}/(\sqrt{\pi} k_f)$. It means that in the limiting case $K \to 0$ the surface temperature at change of the volumes of fluid falls down abruptly to zero value, and then starts to increase monotonically. Let us obtain relations for the following quantities:

- Averaged temperature difference $\langle \vartheta \rangle = 4q\sqrt{\alpha_f \tau_0}/(3\sqrt{\pi} k_f)$
- ATHTC $\langle h \rangle = \sqrt{\pi} k_f / \sqrt{\alpha_f \tau_0}$
- EHTC $h_m = 3\sqrt{\pi} k_f / (4\sqrt{\alpha_f \tau_0})$

An analysis of the transition from the case $K \to \infty$ to $K \to 0$ results in the following conclusions:

- Despite the radical reorganization of the temperature field, oscillations in a body, the EHTC and the ATHTC differ from each other insignificantly (no more than by 25%). Even though this fact is unexpected, it agrees with the physically natural (in other words, physically expected) way of the thermal effect of a wall.
- The EHTC not only does not decrease, but also, on the contrary, increases by $\approx 18\%$. This result is completely unexpected. The reason of this metamorphosis consists, apparently, in reorganization actually of the ATHTC: for the case of $q = $ const, it appears to be $\pi/2$ times higher, than for the case of $\vartheta = $ const.
- An uncontroversial conclusion follows from the above-mentioned limiting estimations that there is practically no effect of the thermal conjugation within the framework of the model of periodic contacts. More precisely: this effect is so weak that it is not visible on the background of the changes in the character of oscillations of the THTC. This discouraging circumstance can induce quite a critical analysis of applications of the model of periodic contacts available in the literature.

For example, in [17] a method of calculation of the influence of a solid body's material on the growth rate of a steam bubble over a heated surface is proposed for the nucleate boiling regime in a liquid. The method is based on the model of a one-time thermal contact. It is implicitly supposed in this method that, after the termination of the interaction, both volumes (liquid and solid) are replaced with new ones. As a matter of fact, this means the replacement of a periodic problem considered above by a problem of one-time thermal contact of two media with homogeneous initial distributions of temperatures. As it is known [5], a solution of latter can be written in the following simple from

$$\frac{\vartheta}{\vartheta_0} = \frac{K}{1+K}, \qquad (1.29)$$

where ϑ_0 is a difference of temperatures between the isothermal bodies before their contact, ϑ is the temperature difference for a fluid volume after the contact. One should point out that, at the given problem statement, the condition $\vartheta = \text{const}$ holds for the entire period of contact. Thus, in the model of one-time thermal contact "conductive condition of periodicity" is absent, with a completely new pair "fluid – solid body" being used for a description of each new contact. The confusion and misunderstanding arising as a result of this in determining of the average temperatures and heat fluxes on a heat transfer surface makes this model incorrect. Apparently, (1.29) has laid a foundation of the correlation from [14], providing introduction of an empirical correction factor such as $\sqrt{K}$ in the formula for a stationary HTC at nucleate boiling.

At the same time, noticeable influence of the complex $\sqrt{kc\rho}$ (coefficient of the thermal activity of a wall) on the measured HTC at nucleate boiling of a liquid is an experimentally established fact. So, it was found in experiments [18] that replacing the heater's material from copper to stainless steel results in a decrease in the heat transfer intensity at boiling cryogenic liquids by an order of magnitude: ≈ 12 times at boiling of nitrogen and ≈ 40 times at boiling of helium. Therefore, there is an open question in front of the theory of nucleate boiling to search for the correct models describing thermal influence of a wall on the average intensity of heat transfer.

1.5 Hydrodynamic HTC

As it was mentioned above, an exact specification of all parameters of the THTC is possible only in view of the exact knowledge of all fields of velocities and temperatures for each particular process with allowance for temperature conjugation between the flowing fluid and the body. Such situation can take place as a result of either (a) a global solution of the system of the unsteady differential equations for the substance transfer in the contacting media or (b) a global experiment, which has been carried out with the help of an ideal instrumentation measuring fields of temperatures and heat fluxes in the coolant and in the wall. Acquisition of the full information for the real unsteady (stochastic) process is believed to be unreal, owing to well-known difficulties in mathematical solution and measuring techniques. For the overwhelming majority of applications, however, so detailed information on fluctuating fields of actual parameters is redundantly detailed and superfluous. Therefore, use of the THTC "specified from the outside" cardinally simplifies this situation: an initial conjugate problem for a system "coolant – wall" is replaced by a boundary value problem for the heat conduction equation in the wall. Thus there is an opportunity to obtain analytical solutions for a series of interesting and actual cases in the applications for the EHTC. It is especially significant in that sense that the structure of real processes, as a rule, is defined by simultaneous influence of many factors. Therefore, direct numerical solutions of a particular problem will inevitably reflect only some special cases of the general multiparameter problem. For the determination of the EHTC, we shall attribute a characteristic (typical) function $h(Z,\tau)$, i.e., a "hydrodynamically determined THTC," to each considered process. As shown above, for a series of processes (such as slug flow of a two-phase medium, wave flow of liquid films, a pulsing jet flow over a body, near-wall turbulent flows), a correct definition of the THTC "from hydrodynamic reasons" is physically quite justifiable. An important specificity of the considered processes consists, thus, in an opportunity of a solution of the heat conduction equation for a wall with "an externally specified" (independent of the thermal influence of a wall) BC of the third kind.

A considerably more complex case of thermal interface is represented by the process of nucleate boiling. As it is known [19], heat transfer intensity at boiling is determined by such factors as velocities and the periods of growth of steam bubbles, density of the bubble-producing cites, a temperature head at the beginning of boiling (superheating) etc. These characteristics generally depend on thermophysical properties and thickness of a heat transferring wall, and in some cases (for example, at nucleate boiling of liquids) effects of this influence can be rather significant. Hence, the THTC describing the process of nucleate boiling also should depend on parameters of conjugation.

It is necessary to emphasize that the method developed in the present work and based on the use of the THTC does not depend on the type of functions $h(Z,\tau)$ and is universal in this sense. However, from the point of view of a practical use of this method, a method of specification of the THTC is important. As shown above, information on the hydrodynamic structure of the flow is sufficient for this purpose in some cases. In this case, a replacement of one heat transferring wall by another (made of a different material, having different thickness, heat input conditions), with a two-layer plate or a body of different geometry, etc. does not result in a change of the behavior of the THTC. Then, having solved the heat conduction equation for various bodies with a BC of the third kind, it is possible to obtain a certain "set" of values of the EHTC. Distinction of these EHTC-functions among themselves will also express qualitative and quantitative effects of thermal influence of a body on the averaged heat transfer intensity. For the case of nucleate boiling, a change of the conjugation parameters should result also in a change of the actual THTC. A remedy here can consist, apparently, in a development of initial theoretical models for the THTC, taking into account initial influence on them of the conjugation parameters. Then our method can allow introducing correctly additional amendments to such parameters taking onto account the effect of conjugation. One should also note, that a physical class of the heat transfer processes with the periodic intensity including "hydrodynamically determined" THTC is rather wide and covers, apparently, overwhelming majority of the engineering applications. This circumstance is a powerful argument in favor of the actuality of the present research.

1.6 Previous Investigations of Heat Transfer Processes with Periodic Intensity

Experimental and numerical investigations of heat transfer at laminar flow in a pipe under conditions of periodic oscillations of pressure were carried out in [20, 21]. Similar studies applicable to a flow of gas in regenerators under conditions of an intermittently reversed mass flow-rate have been carried out in [22, 23]. These works based their analysis on a nonconjugate problem statement, i.e., used an initially set wall temperature (fixed value). One should notice that this fact is quite justifiable for the conditions of those particular experiments. It is clear that for the use of air as the coolant, treatment of a physical problem in a thermally conjugate statement is practically unnecessary. Thermophysical properties are many times less than those of metals, and consequently gases cannot basically render any appreciable influence on the temperature field of in a body. On the other hand, the interesting experimental and theoretical information on local HTC periodically changing in time obtained in [20–23] makes a valuable database for a computation of parameters under conditions of the hydrodynamically determined HTC. An indirect confirmation of the presence of the thermal influence of a solid body was obtained in [24]. An experimental research of temperature oscillations in a wall for turbulent flow of water in a channel performed in this work has shown that these oscillations appear for a case of the wall made of stainless steel and are practically completely absent for the case of copper wall. The class of conjugate stationary problems of heat transfer in a laminar boundary layer has been analytically investigated in a series of works by Dorfman [25–27]. An important achievement of the specified works is the substantiation of generalization of the self-similar variables proposed by Blasius [15] and their further use for the case of thermal conjugation. Later analytical solutions of the stationary conjugate problems have been obtained at flow of liquids in channels using a similar approach [28, 29]. The authors of [30, 31] have numerically investigated a stationary conjugate problem for a flow in a channel with discrete sources of heating. It represents an important step on studying of spatioperiodic type of the thermal conjugation. However, in the specified works there is no generalization given concerning the results of the investigated thermal influence.

1.7 Analytical Methods

As known, the majority of problems of hydrodynamics and heat transfer are described by partial differential equations. So, Navier–Stokes and energy equations represent quasilinear partial differential equations, which solution in most cases can only be obtained with the help of numerical methods. This can lead to a "natural" conclusion about an absolute priority of numerical solutions in the specified area of research. However, analytical solutions of the fluid flow and heat transfer problems play a significant role even in the current computer age. They possess the following decisive advantages in comparison with numerical methods:

- The value of the analytical approach consists in an opportunity of the closed qualitative description of the process, revealing of the full list of dimensionless characteristic parameters and their hierarchical classification based on the criteria of their importance.
- Analytical solutions possess a necessary generality, so that a variation of boundary and inlet conditions allows carrying out parametrical investigations.
- In order to validate numerical solutions of the full differential equations, it is necessary to have basic (often rather simple) analytical solutions of the equations for some obviously simplified cases (after an estimation and omission of negligible terms).
- In a global aspect, an analytical solution can be used for a direct validation of the correctness in the statement of numerical investigations applicable to a particular problem.

Analytical investigations of a wide spectrum of fluid flow and heat transfer problems have been carried out in the book of Weigand [32]. Parabolic, elliptic, and hyperbolic partial differential equations of second order were considered. Solutions of a wide class of problems with the help of the classical method of separation of variables are also presented in the book. Classical and modern methods of the analytical solution of the hydrodynamics and heat transfer problems are considered for flow of a fluid in a channel for various conditions: stationary and unsteady (including periodically fluctuating) flow, flow over a thermal initial length, flow in an axially rotating pipe. A limiting case of large eigenvalues of the solution is considered, as well as asymptotic solutions for small Peclet numbers. The class of nonlinear differential equations, opportunities of their linearization, application of self-similar variables have been also thoroughly investigated. The value of the book of Weigand [32] in the sense of the method proposed in the present work consists in the availability of a representative database for determination of "hydrodynamically determined HTC," i.e., in the formation of a theoretical basis for calculating the EHTC. The present book overviewes and generalizes from a single viewpoint results published by the author in works [33–68].

References

1. Hinze JO (1975) Turbulence. McGraw-Hill, New York.
2. Dietz C, Henze M, Neumann SO, von Wolfersdorf J, Weigand B (2005) Numerical and experimental investigation of heat transfer and fluid flow around a vortex generator using explicit algebraic models for the turbulent heat flux.

Proc. of the 17th Int. Symp. on Airbreathing Engines, September 4–9, Munich, Germany, Paper ISABE-2005-1197.
3. Baehr HD, Stephan J (1998) Heat and Mass Transfer. Springer, Berlin Heidelberg New York.
4. Cebeci T (2002) Convective Heat Transfer. Springer, Berlin Heidelberg New York.
5. Carslaw HS, Jaeger JC (1992) Conduction of Heat in Solids. Clarendon Press, London, Oxford.
6. Wallis GB (1969) One-Dimensional Two-phase Flow, McGraw-Hill, New York.
7. Mayinger F (1982) Strömung und Wärmeübergang in Gas-Flüssigkeits-Gemischen. Springer, Wien, New York.
8. Kapitsa PL (1948) Wave flow of thin layers of a viscous liquid. Part 1. Free flow. Zh Eksp Teor Fiz 18 (1): 1–28 (in Russian).
9. Kapitsa PL, Kapitsa SP (1949) Wave flow of thin layers of a viscous liquid. Part II. Fluid flow in the presence of continuous gas flow and heat transfer. Zh Eksp Teor Fiz 19 (2): 105–120 (in Russian).
10. Corino ER, Brodkey RS (1969) A visual investigation of the wall region in turbulent flow. J Fluid Mech 37 (1):1–30.
11. Kim HT, Kline SJ, Reynolds WC (1971) The production of turbulence near a smooth wall in a turbulent boundary layer. J Fluid Mech 50 (1): 133–160.
12. Reyer V (2002) Ein Verfahren zur simultanen Berechnung gekoppelter transienter Temperaturfelder in Strömungen und Strukturen. Dissertation, Berlin Technical University.
13. Webster RS (2001) A numerical study of the conjugate conduction–convection heat transfer problem. Dissertation, Michigan State University.
14. Gorenflo D (2002) Behältersieden (Sieden bei freier Konvektion). VDI – Wärmeatlas, Hab. Springer, Berlin Heidelberg New York.
15. Schlichting H, Gersten K (1997) Grenzschicht-Theorie. Springer, Berlin Heidelberg New York.
16. Cebeci T, Bradshaw P (1984) Physical and Computational Aspects of Convective Heat Transfer. Springer. Berlin Heidelberg New York.
17. Ametistov EV, Grigoriev VA, Pavlov YM (1972) Effect of thermophysical properties of heating surface material on heat transfer during boiling of water and ethanol. High Temp 10: 821–823.
18. Grigoriev VA, Pavlov YM, Ametisov EV, Klimenko AV, Klimenko VV (1977) Concerning the influence of thermal properties of heating surface material on heat transfer intensity of nucleate pool boiling of liquids including cryogenic ones. Cryogenics 2: 94–96.
19. Stephan K (1992) Heat Transfer in Condensation and Boiling. Springer, Berlin Heidelberg New York.
20. Habib MA, Attya AM, Said SAM, Eid AI, Aly AZ (2004) Heat transfer characteristics and Nusselt number correlation of turbulent pulsating pipe air flows. Heat Mass Transf 40: 307–318.
21. Yakhot A, Arad M, Ben-Dor G (1999) Numerical investigation of a laminar pulsating flow in a rectangular duct. Int J Numer Methods Fluids 29: 935–950.
22. Walther Ch, Kühl H.-D, Pfeffer Th, Schulz S (1998) Influence of developing flow on the heat transfer in laminar oscillating pipe flow. Forschung im Ingenieurwesen 64: 55–64.
23. Walther C, Kühl H-D, Schulz S (2000) Numerical investigations on the heat transfer in turbulent oscillating pipe flow. Heat Mass Transf 36: 135–141.

24. Mosyak A, Pogrebnyak E, Hetsroni G (2001) Effect of constant heat flux boundary condition on wall temperature fluctuations. ASME J Heat Transf 123: 213–218.
25. Dorfman AS (1985) A new type of boundary condition in convective heat transfer problems. Int J Heat Mass Transf 28: 1197–1203.
26. Dolinskiy AA, Dorfman AS, Davydenko BV (1989) Conjugate heat and mass transfer in continuous processes of convective drying. Int J Heat Mass Transf 34: 2883–2889.
27. Dorfman AS (2004) Transient heat transfer between a semi-infinite hot plate and a flowing cooling liquid film. ASME J Heat Transf 126: 149–154.
28. Kiwan SM, Al-Nimr MA (2002) Analytical solution for conjugated heat transfer in pipes and ducts. Heat Mass Transf 38: 513–516.
29. Soliman HM, Rahman MM (2006) Analytical solution of conjugate heat transfer and optimum configurations of flat-plate heat exchangers with circular flow channels. Heat Mass Transf 42: 596–607.
30. Wang Q, Jaluria Y (2004) Three-dimensional conjugate heat transfer in a horizontal channel with discrete heating. ASME J Heat Transf 126: 642–647.
31. Weigand B, Lauffer D (2004) The extended Graetz problem with piecewise constant wall temperature for pipe and channel flows. Int J Heat Mass Transf 471: 5303–5312.
32. Weigand B (2004) Analytical Methods for Heat Transfer and Fluid Flow Problems. Springer, Berlin Heidelberg New York.
33. Labuntsov DA, Zudin YB (1977) Peculiarities of the process of heat transfer from a surface of a plate to a flow with a spatiotemporal periodic variation of the heat transfer coefficient. Part 1. General analysis. Works of Moscow Power Engineering Institute. Issue 347: 84–92 (in Russian).
34. Labuntsov DA, Zudin YB (1977) Peculiarities of the process of heat transfer from a surface of a plate to a flow with a spatiotemporal periodic variation of the heat transfer coefficient. Part 2. Solution of characteristic problems. Works of Moscow Power Engineering Institute. Issue 347: 93–100 (in Russian).
35. Zudin YB, Labuntsov DA (1978) Peculiarities of heat transfer at periodic asymmetrical regime. Works of Moscow Power Engineering Institute. Issue 377: 35–39 (in Russian).
36. Zudin YB (1980) Analysis of Heat-Transfer Processes of Periodic Intensity. Dissertation. Moscow Power Engineering Institute (in Russian).
37. Labuntsov DA, Zudin YB (1984) Heat-Transfer Processes of Periodic Intensity, Energoatomizdat, Moscow (in Russian).
38. Zudin YB (1991) Calculation of an empirical heat-transfer coefficient with a stepped periodic change in heat-transfer rate. High Temp 29: 740–745.
39. Zudin YB (1991) A method of heat-exchange calculation in the presence of periodic intensity fluctuations. High Temp 29: 921–928.
40. Zudin YB (1992) Analog of the Rayleigh equation for the problem of bubble dynamics in a tube. J Eng Phys Thermophys 63: 672–675.
41. Zudin YB (1993) The calculation of parameters of the evaporating meniscus a thin liquid film. High Temp 31: 714–716.
42. Zudin YB (1994) Calculation of effect for supplying heat to the wall on the averaged heat exchange coefficient. Thermophys Aeromech 1: 117–119.
43. Zudin YB (1995) Averaged heat transfer during periodic fluctuations of the heat transfer intensity of the surface of a plate, a cylinder, or a sphere. J Eng Phys Thermophys 68: 193–196.

44. Zudin YB (1995) Calculation of heat transfer characteristics with periodic pulsations of "cellular structure" intensity. Appl Energy Russ J Fuel Power Heat Syst 33: 151–159.
45. Zudin YB (1995) Design of the wall heat effect on averaged convective heat transfer in processes of heat exchange with periodic intensity. Appl Energy Russ J Fuel Power Heat Syst 33: 76–81.
46. Zudin YB (1995) Averaged heat exchange for double-sided periodicity. Thermophys Aeromech 2: 281–287.
47. Zudin YB (1996) On two types of pulsations of true heat transfer coefficient (a progressive wave and a cell). Thermophys Aeromech 3: 341–346.
48. Zudin YB (1996) Pulse law of true heat transfer coefficient pulsations. Appl Energy Russ J Fuel Power Heat Syst 34: 142–147.
49. Zudin YB (1996) Theory on heat-transfer processes of periodic intensity Habilitation. Moscow Power Engineering Institute (in Russian).
50. Zudin YB (1997) Calculation of critical thermal loads under extreme intensities of mass forces. Heat Transf Res 28: 481–483.
51. Zudin YB (1997) Influence of the coefficient of thermal activity of a wall on heat transfer in transient boiling. J Eng Phys Thermophys 71: 696–698.
52. Zudin YB (1997) Law of vapor-bubble growth in a tube in the region of low pressures. J Eng Phys Thermophys 70: 714–717.
53. Zudin YB (1997) The use of the model of evaporating macrolayer for determining the characteristics of nucleate boiling. High Temp 35: 565–571.
54. Zudin YB (1998) Calculation of the surface density of nucleation sites in nucleate boiling of a liquid. J Eng Phys Thermophys 71: 178–183.
55. Zudin YB (1998) Boiling of liquid in the cell of a jet printer. J Eng Phys Thermophys 71: 217–220.
56. Zudin YB (1998) Effect of the thermophysical properties of the wall on the heat transfer coefficient. Therm Eng 45 (3): 206–209.
57. Zudin YB (1998) The distance between nucleate boiling sites. High Temp 36: 662–663.
58. Zudin YB (1998) Temperature waves on a wall surface. Russ Dokl Phys J Acad Sci 43 (5) 313–314.
59. Zudin YB (1999) Burn-out of a liquid under conditions of natural convection. J Eng Phys Thermophys 72: 50–53.
60. Zudin YB (1999) Wall non-isothermicity effect on the heat exchange in jet reflux. J Eng Phys Thermophys 72: 309–312.
61. Zudin YB (1999) Model of heat transfer in bubble boiling. J Eng Phys Thermophys 72: 438–444.
62. Zudin YB (1999) Self-oscillating process of heat exchange with periodic intensity. J Eng Phys Thermophys 72: 635–641.
63. Zudin YB (1999) The effect of the method for supplying heat to the wall on the averaged heat-transfer coefficient in periodic rate heat-transfer prozesses. Therm Eng 46 (3): 239–243.
64. Zudin YB (1999) Harmonic law of fluctuations of the true heat transfer coefficient. Thermophys Aeromech 6: 79–88.
65. Zudin YB (1999) Some properties of the solution of the heat-conduction equation with periodic boundary condition of third kind. Thermophys Aeromech 6: 391–398.

66. Zudin YB (2000) Processes of heat exchange with periodic intensity. Therm Eng 47 (6): 124–128.
67. Zudin YB (2000) Analysis of the processes of heat transfer with periodic intensity with allowance for temperature fluctuations in the heat carrier. J Eng Phys Thermophys 73: 243–247.
68. Zudin YB (2000) Averaging of the heat-transfer coefficient in the processes of heat exchange with periodic intensity. J Eng Phys Thermophys 73: 643–647.

2

Construction of a General Solution of the Problem

2.1 Boundary Value Problem for the Heat Conduction Equation

Let us carry out an analytical consideration of a boundary value problem for the two-dimensional transient heat conduction equation [1–4]. An object of research is a flat plate of the thickness δ, Fig. 2.1. In accordance with the basic idea of the present research, on an internal surface of a plate ($X = \delta$) a boundary condition (BC) of the third kind is used:

$$h(z, \tau) = h(\xi). \tag{2.1}$$

Here $\xi = \tau/\tau_0 \mp Z/Z_0 = t \mp z$ is the generalized coordinate of the progressive wave developing from left to right with the coordinate Z along the heat transfer surface; Z_0, τ_0 are the spatial and time periods of oscillations, respectively. Let us, for clarity, agree to name boundary conditions on an external surface of a body (at $X = 0$) a *Thermal Boundary Conditions* (TBC). In general, TBC of the following three kinds can be used:

(a) Constant temperature

$$\vartheta_0 = \text{const.} \tag{2.2a}$$

(b) Constant heat flux density

$$q_0 = \text{const.} \tag{2.2b}$$

(c) An adiabatic surface, constant volumetric heat sources

$$q_0 = 0, q_V = \text{const.} \tag{2.2c}$$

The heat conduction equation for the considered general case of spatiotemporal oscillations is given by [5, 6]

$$c\rho \frac{\partial \vartheta}{\partial \tau} = k\left(\frac{\partial^2 \vartheta}{\partial X^2} + \frac{\partial^2 \vartheta}{\partial Z^2}\right) + q_V. \tag{2.3}$$

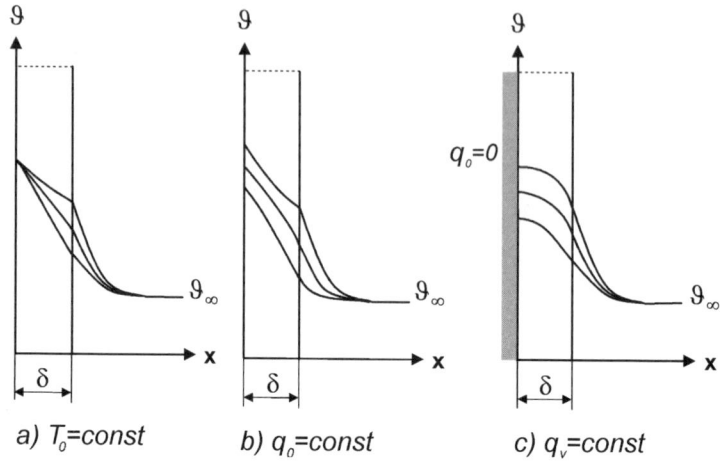

Fig. 2.1. Schematic of the heat transfer process with periodic intensity

Owing to the linearity of (2.3), its solution can always be presented as a superposition: $\vartheta = \langle\vartheta\rangle(X) + \hat{\vartheta}(X, Z, \tau)$. The stationary component satisfies the equation

$$k\frac{d^2\langle\vartheta\rangle}{dX^2} + q_V = 0 \qquad (2.4)$$

and the corresponding TBC. The oscillation component is described by the equation

$$c\rho\frac{\partial\hat{\vartheta}}{\partial\tau} = k\left(\frac{\partial^2\hat{\vartheta}}{\partial X^2} + \frac{\partial^2\hat{\vartheta}}{\partial Z^2}\right). \qquad (2.5)$$

The solution of the stationary equation is given by

$$\vartheta_0 = \text{const} : \langle\vartheta\rangle = \langle\vartheta_0\rangle - \langle\vartheta_\infty\rangle - AX, \qquad (2.6\text{a})$$

$$q_0 = \text{const} : \langle\vartheta\rangle = B - \frac{q_0 X}{k}, \qquad (2.6\text{b})$$

$$q_V = \text{const}\,(q_0 = 0) : \langle\vartheta\rangle = C - \frac{q_V X^2}{2k}. \qquad (2.6\text{c})$$

Here A, B, C are constants. The oscillation component of a temperature field depends on the generalized coordinate of a progressive wave ξ and a spanwise coordinate X: $\hat{\vartheta}(X, Z, \tau) = \hat{\vartheta}(X, \xi,)$. Therefore it is possible to rewrite (2.5) after some simple transformations as

$$m\frac{\partial\hat{\vartheta}}{\partial\xi} - \frac{\partial^2\hat{\vartheta}}{\partial\xi^2} = \frac{\partial^2\hat{\vartheta}}{\partial x^2}, \qquad (2.7)$$

where $x = X/Z_0$ is the dimensionless spanwise coordinate. There exist three possible variants of the TBC for (2.7)

2.1 Boundary Value Problem for the Heat Conduction Equation

$$\vartheta_0 = \text{const}: \quad \hat{\vartheta}_0 = 0, \qquad (2.8a)$$

$$q_0 = \text{const}: \quad \hat{\vartheta}_0^{\bullet} = 0, \qquad (2.8b)$$

$$q_0 = 0 \, (q_V = \text{const}): \quad \hat{\vartheta}_0^{\bullet} = 0. \qquad (2.8c)$$

Here $\hat{\vartheta}_0^{\bullet} = \left.\frac{\partial \hat{\vartheta}}{\partial X}\right|_{X=0}$ is a gradient of the oscillating temperature at $X = 0$. It follows from (2.8b) and (2.8c) that the last two cases are identical. Therefore we shall everywhere distinguish only two alternative TBC: $\vartheta_0 = \text{const}$ and $q_0 = \text{const}$ Periodic solutions of (2.7) satisfying the BC (2.8) have the following form [7]:

(a) TBC of $\vartheta_0 = \text{const}$

$$\hat{\vartheta} = \sum_{n=1}^{\infty} \left\{ A_n \frac{\text{sh}\left[(r_n + is_n)\, x\right]}{\text{sh}\left[(r_n + is_n)\right]\bar{\delta}} \exp(in\xi) + A_n^* \frac{\text{sh}\left[(r_n - is_n)\, x\right]}{\text{sh}\left[(r_n - is_n)\bar{\delta}\right]} \exp(-in\xi) \right\}. \qquad (2.9a)$$

(b) TBC of $q_0 = \text{const}$

$$\hat{\vartheta} = \sum_{n=1}^{\infty} \left\{ A_n \frac{\text{ch}\left[(r_n + is_n)\, x\right]}{\text{ch}\left[(r_n + is_n)\right]\bar{\delta}} \exp(in\xi) + A_n^* \frac{\text{ch}\left[(r_n - is_n)\, x\right]}{\text{ch}\left[(r_n - is_n)\bar{\delta}\right]} \exp(-in\xi) \right\}. \qquad (2.9b)$$

Here $r_n = \frac{n}{\sqrt{2}}\left[\sqrt{1 + \left(\frac{m}{n}\right)^2} + 1\right]^{1/2}$, $s_n = \frac{n}{\sqrt{2}}\left[\sqrt{1 + \left(\frac{m}{n}\right)^2} - 1\right]^{1/2}$; $x, \bar{\delta}$ are dimensionless values of the spanwise coordinate and wall thickness, respectively; A_n, A_n^* are complex conjugate eigenvalues of the considered boundary problem. The eigenvalues should be determined from the BC at $X = \delta$

$$h\left(\xi\right)\vartheta_\delta = -k \left.\frac{\partial \vartheta}{\partial X}\right|_{X=\delta}. \qquad (2.10)$$

The true heat transfer coefficient (THTC) is represented as a Fourier series

$$h\left(\xi\right) = \langle h \rangle \left\{ 1 + \sum_{n=1}^{\infty} \left[C_n \exp(in\xi) + C_n^* \exp(-in\xi)\right] \right\}. \qquad (2.11)$$

Thus, as it has already been mentioned before, the parameters $\langle h \rangle, C_n, C_n^*$ are considered as a priori known. After the substitution of (2.9, 2.11) into the BC (2.10) and the determination of the values A_n, A_n^*, the temperature field in a wall is known. Then, it is possible to calculate the average difference of temperatures "body – fluid" $\langle \vartheta_\delta \rangle$ and the average heat flux density $\langle q_\delta \rangle$, which penetrates through a heat transfer surface [8]. In a result it is possible to determine required experimental heat transfer coefficient (EHTC)

$$h_m = \frac{\langle q_\delta \rangle}{\langle \vartheta_\delta \rangle}. \qquad (2.12)$$

Thus, at s specified THTC, determined by fluid mechanics, the solution of a problem gives EHTC, dependent on the properties of the body. The investigation of this dependence takes the central place in the present book. For the subsequent analysis, it is expedient to introduce a dimensionless relative parameter, i.e., ratio of the EHTC to ATHTC

$$\varepsilon = \frac{h_m}{\langle h \rangle}. \tag{2.13}$$

This parameter which reflects the quantitative effect of the interface "fluid flow – body" will be called the *factor of conjugation* (FC).

2.2 Spatial and Temporal Types of Oscillations

An important parameter of a problem determining the type of oscillations (spatial or temporal), is the value of $m = Z_0^2/(\alpha \tau_0)$, i.e., the inverse Fourier number. A limiting case of $m \to 0$ corresponds to an unlimited extension of the time period: $\tau_0 \to \infty$. Thus the progressive wave of oscillations is frozen ($r_n \to n, s_n \to 0, \xi \to z = Z/Z_0$), and the THTC changes along the heat transfer surface under the spatial periodic law $h(\xi) \to h(z)$. Equations (2.9a) and (2.9b) thus take a simpler form

(a) at $\vartheta_0 = $ const

$$\hat{\vartheta} = \sum_{n=1}^{\infty} [A_n \exp(inz) + A_n^* \exp(-inz)] \frac{\text{sh}(nx)}{\text{sh}(n\bar{\delta})}, \tag{2.14a}$$

(b) at $q_0 = $ const

$$\hat{\vartheta} = \sum_{n=1}^{\infty} [A_n \exp(inz) + A_n^* \exp(-inz)] \frac{\text{ch}(nx)}{\text{ch}(n\bar{\delta})}. \tag{2.14b}$$

The limiting case of a stopped progressive wave considered here is described by the two-dimensional stationary heat conduction equation

$$\frac{\partial^2 \hat{\vartheta}}{\partial z^2} + \frac{\partial^2 \hat{\vartheta}}{\partial x^2} = 0. \tag{2.15}$$

The limiting case $m \to \infty$ can be obtained, if the speed of development of a progressive wave along a surface of a body tends to infinity. This will correspond also to an unlimited extension of the spatial period of oscillations: $Z_0 \to \infty, r_n \to \sqrt{mn/2}, s_n \to \sqrt{mn/2}, (r_n \pm is_n)x \to \sqrt{n/2}(1 \pm i)(X/\sqrt{\alpha \tau_0}) = \sqrt{n/2}(1 \pm i)\tilde{x}, \xi \to t = \tau/\tau_0$. Equations (2.9a) and (2.9b) can be rewritten to give

(a) TBC $\vartheta_0 = $ const:

$$\hat{\vartheta} = \sum_{n=1}^{\infty} \left\{ A_n \frac{\sinh\left[\sqrt{n/2}\,(1+i)\,\tilde{x}\right]}{\sinh\left[\sqrt{n/2}\,(1+i)\,\tilde{\delta}\right]} \exp(int) \right.$$

$$\left. + A_n^* \frac{\sinh\left[\sqrt{n/2}\,(1-i)\,\tilde{x}\right]}{\sinh\left[\sqrt{n/2}\,(1-i)\,\tilde{\delta}\right]} \exp(-int) \right\}, \qquad (2.16a)$$

(b) TBC $q_0 = $ const:

$$\hat{\vartheta} = \sum_{n=1}^{\infty} \left\{ A_n \frac{\cosh\left[\sqrt{n/2}\,(1+i)\,\tilde{x}\right]}{\cosh\left[\sqrt{n/2}\,(1+i)\,\tilde{\delta}\right]} \exp(int) \right.$$

$$\left. + A_n^* \frac{\cosh\left[\sqrt{n/2}\,(1-i)\,\tilde{x}\right]}{\cosh\left[\sqrt{n/2}\,(1-i)\,\tilde{\delta}\right]} \exp(-int) \right\}. \qquad (2.16b)$$

This variant is equivalent to a case of synchronous time oscillations of the THTC on the whole surface of $X = \delta$, which is described by the one-dimensional transient heat conduction equation

$$\frac{\partial \hat{\vartheta}}{\partial t} = \frac{\partial^2 \hat{\vartheta}}{\partial \tilde{x}^2}. \qquad (2.17)$$

Unlike in the case of $m \to 0$, the asymptotical solution (2.16) is not so obvious. Really, spatiotemporal character of a progressive wave means that at any speed of its distribution the function $h(\xi)$ for various values of the longitudinal coordinate Z will have different phases. Therefore the limit $m \to \infty$ should be understood as a degeneration of the dependence of temperature oscillations on the spatial period of the oscillations Z_0. Now a natural lengthscale (along the spanwise coordinate of a plate X) becomes the value $\sqrt{\alpha\tau_0}$, i.e., the depth of penetration of a temperature wave into the body.

2.3 Interrelation between the Two Averaged Coefficients of Heat Transfer

The interrelation between the values of $\langle h \rangle$ and h_m plays the central role in the analysis. It is defined from the BC (2.10):

$$h\left(\langle \vartheta_\delta \rangle + \hat{\vartheta}_\delta\right) = -k \frac{\partial \langle \vartheta \rangle}{\partial X}\bigg|_{X=\delta} - k \frac{\partial \hat{\vartheta}}{\partial X}\bigg|_{X=\delta}. \qquad (2.18)$$

2 Construction of a General Solution of the Problem

According to the aforementioned, EHTC is defined as follows:

$$h_m = \frac{\langle q_\delta \rangle}{\langle \vartheta_\delta \rangle} = -\frac{k}{\langle \vartheta_\delta \rangle} \frac{\partial \langle \vartheta \rangle}{\partial X}\bigg|_{X=\delta}. \qquad (2.19)$$

The second (alongside with m) fundamental parameter of the problem is a dimensionless value of the ATHTC (or Biot number – [8])

$$\langle \bar{h} \rangle = \frac{\langle h \rangle Z_0}{k}. \qquad (2.20)$$

Having designated the periodic part of the THTC in (2.11) as

$$\psi = \sum_{n=1}^{\infty} [C_n \exp(in\xi) + C_n^* \exp(-in\xi)], \qquad (2.21)$$

let us rewrite (2.18) as

$$\langle h \rangle (1+\psi)(1+\theta) = h_m - \theta^\bullet \qquad (2.22)$$

with the abbreviations:

$$\theta = \frac{\hat{\vartheta}_\delta}{\langle \vartheta_\delta \rangle}, \theta^\bullet = \frac{1}{\langle \vartheta_\delta \rangle} \frac{\partial \hat{\vartheta}}{\partial x}\bigg|_{x=\bar{\delta}}. \qquad (2.23)$$

The first form of the notation of the BC. Averaging both parts of (2.22) over the period of the oscillations, we obtain a ratio for the EHTC:

$$h_m = \langle h \rangle (1 + \langle \psi\theta \rangle). \qquad (2.24)$$

The ratio for FC follows from the (2.13)

$$\varepsilon = 1 + \langle \psi\theta \rangle. \qquad (2.25)$$

Equation (2.25) allows making important conclusions about the character of the thermal influence of a solid body. We shall consider for clarity a case when the heat is removed from a heat transfer surface by a flowing fluid (corresponding, for example, to a case of cooling of a wall by a boiling liquid). For that part of the period, when the level of heat transfer intensity is above average (active heat transfer), one has $\psi \geq 0$. It follows from physical reasons that the surface temperature in this case will be lower than the average: $\theta \leq 0$. For the other part of the period, when the level of the heat transfer intensity is lower than average (passive heat transfer, $\psi \leq 0$), it is possible to assume from the same reasons, that $\theta \geq 0$. Thus, for the entire period of oscillations one can write

$$\psi \geq 0 \Rightarrow \theta \leq 0, \psi \leq 0 \Rightarrow \theta \geq 0. \qquad (2.26)$$

2.3 Interrelation between the Two Averaged Coefficients of Heat Transfer

This results in the following inequality:

$$\langle \psi \theta \rangle \leq 0, \tag{2.27}$$

which together with (2.25) means

$$\varepsilon \leq 1. \tag{2.28}$$

Equation (2.28), which validity can be strictly proved for the general case (see Appendix A), plays a fundamental role in the present analysis. It means that EHTC (determined in a traditional experiment and used in applied calculations) always it is less ATHTC (determined from the theory of stationary convective heat transfer). At $\theta \to 0$ it follows from (2.25) that $\varepsilon = 1, h_m = \langle h \rangle$. It is clear from physical considerations that at the presence of external hydrodynamical oscillations a degeneration of the temperature oscillations in a wall should take place at an unlimited increase in heat conduction within a body. *The second form of the notation of the BC.* Let us perform some transformations of the notation of the analyzed BC at $x = \bar{\delta}$. Dividing both parts of (2.22) by $\langle \bar{h} \rangle$ one obtains:

$$(1+\psi)(1+\theta) = \varepsilon - \frac{\theta^\bullet}{\langle \bar{h} \rangle}. \tag{2.29}$$

In turn, having divided both parts (2.29) by $1+\psi$, one obtains:

$$(1+\theta) = \frac{\varepsilon}{(1+\psi)} - \frac{\theta^\bullet}{(1+\psi)\langle \bar{h} \rangle}. \tag{2.30}$$

Averaging of (2.29) gives a ratio expressed as (2.25) that is already known to us. In turn, averaging of (2.30) gives an alternative form of a notation of the BC

$$\varepsilon = \left\langle \frac{1}{1+\psi} \right\rangle^{-1} \left(1 + \frac{1}{\langle \bar{h} \rangle} \left\langle \frac{\theta^\bullet}{1+\psi} \right\rangle \right). \tag{2.31}$$

Equation (2.31) allows determining a minimal value of the FC corresponding to the maximal effect of thermal influence of a solid body. It is obvious from the physical considerations that for that part of the period, when the level of heat transfer intensity is above the average (active heat transfer), the oscillation of the heat flux density is positive. In view of Fourier's law, this effectively means: $\psi \geq 0, \theta^\bullet \leq 0$. For that part of the period, when the level of the heat transfer intensity is lower than the average (passive heat transfer), the same reasoning gives $\psi \leq 0, \theta^\bullet \geq 0$. The above-stated estimations can be rewritten in the following notation

$$\frac{1}{1+\psi} \geq 1 \Rightarrow \theta^\bullet \leq 0, \quad \frac{1}{1+\psi} \leq 1 \Rightarrow \theta^\bullet \geq 0. \tag{2.32}$$

Hence, the absolute value of $\theta^\bullet/(1+\psi)$ for the period of active heat transfer $(\theta^\bullet/(1+\psi) \leq 0)$ is less than that for the period of passive heat transfer

$((\theta^{\bullet})/(1+\psi) \geq 0)$. As a result, on the average for the entire period of heat transfer we can find that

$$\left\langle \frac{\theta^{\bullet}}{1+\psi} \right\rangle \geq 0. \tag{2.33}$$

It follows from here with the help of (2.31)

$$\varepsilon \geq \left\langle \frac{1}{1+\psi} \right\rangle^{-1}. \tag{2.34}$$

The inequality (2.34), which validity is also mathematically proven in Appendix A, shows, that the minimally possible value of FC is defined by the type of a periodic component (or, more exactly saying, by the amplitude of oscillations) of the THTC. It follows from physical reasoning that the equality $\varepsilon = \langle 1/(1+\psi) \rangle^{-1}$ can be fulfilled in a hypothetical case of a solid body with zero thermal conductivity when the heat flux oscillations in it degenerate.

It is convenient to rewrite both fundamental inequalities, namely (2.28) and (2.34), in the form of (2.35):

$$\left\langle \frac{1}{1+\psi} \right\rangle^{-1} \leq \varepsilon \leq 1, \tag{2.35}$$

determining a range of variation of the FC. During the further analysis, both equivalent forms of notation of the BCs (2.29) and (2.30) will be used. It is important to note that asymptotic behavior of the FC can be investigated directly from the relations (2.29) and (2.30), i.e., before the solution of a boundary problem for the heat conduction equation in a body.

2.4 Dimensionless Parameters

Consideration of the heat conduction equation (2.7) and the BC (2.29) at $X = \delta$ (or its equivalent in the form of (2.30)) shows that the FC generally depends on the following dimensionless parameters:

- $m = \frac{Z_0^2}{\alpha \tau_0}$ that is a ratio of the spatial and temporal periods of oscillations (an inverse Fourier number)
- $\langle \bar{h} \rangle = \frac{\langle h \rangle Z_0}{k}$, dimensionless averaged true heat transfer coefficient (ATHTC), or Biot number
- $\bar{\delta} = \frac{\delta}{Z_0}$, dimensionless wall thickness (a flat plate)
- $\psi(\xi)$, that means the type and the amplitude of oscillations of the THTC

For a limiting case of only spatial oscillations $m \to 0$, $\psi(\xi) \to \psi(z)$ two key parameters $\langle \bar{h} \rangle, \bar{\delta}$ preserve their initial form. In a limiting case of only temporal oscillations $-m \to \infty$, $\psi(\xi) \to \psi(t)$, a redefinition of the lengthscale takes place $Z_0 \Rightarrow \sqrt{\alpha \tau_0}$. This will lead also to the redefinition of the two basic dimensionless parameters:

- $\langle \tilde{h} \rangle = \frac{\langle h \rangle \sqrt{\alpha \tau_0}}{k}$ that is the Biot number
- $\tilde{\delta} = \frac{\delta}{\sqrt{\alpha \tau_0}}$, i.e., dimensionless wall thickness

In view of the two possible alternative TBCs [(2.8a) and (2.8b)], the number of determining parameters is actually doubled. Thus, the considered problem becomes essentially multiparametrical and includes a plenty of various practically important individual problems, which can considerably differ in quantitative and qualitative aspects.

2.5 Factor of Conjugation: An Analysis of Limiting Variants

Asymptotical solution $\langle \bar{h} \rangle \to 0, \langle \tilde{h} \rangle \to 0$. This limiting case corresponds to an infinitely large thermal conductivity of a solid body ($k \to \infty$). As the oscillations of heat flux in a heat transferring body $\hat{q} \sim k\hat{\vartheta}/X$ should be limited, temperature oscillations $\hat{\vartheta}$ in the whole body will be negligibly small. At $\theta \to 0$, one can derive from (2.29):

$$\varepsilon \to 1, \tag{2.36}$$

i.e., $h_m \to \langle h \rangle$. Thus, the thermal conjugation degenerates in the asymptotic case

$$\langle \bar{h} \rangle \to 0.$$

Asymptotical solution $\langle \bar{h} \rangle \to \infty, \langle \tilde{h} \rangle \to \infty$. This asymptotical solution is opposite to the previous one and describes a limiting case with negligible small thermal conductivity of a body ($k \to 0$). It is physically obvious, that temperature oscillations θ and their gradients $\theta^{\bullet}$ on a heat transfer surface will be limited. Then it follows from (2.30) at $\langle \bar{h} \rangle \to \infty$

$$\varepsilon \to \varepsilon_{\min} = \left\langle \frac{1}{1+\psi} \right\rangle^{-1}. \tag{2.37}$$

This means that the FC achieves the minimally possible value determined by the type of the periodic function $\psi(\xi)$. Considered asymptotical solution (that characterizes the limiting effect of the thermal influence of a solid body) is of a significant interest. We shall illustrate this by the example of several particular functions of $\psi(\xi)$: cosine function

$$h = \langle h \rangle [1 + b\cos(\xi)], \varepsilon_{\min} = \sqrt{1-b^2}, \tag{2.37a}$$

and step function

$$\left. \begin{array}{l} 0 \leq \xi \leq \pi : h = \langle h \rangle (1+b) \\ \pi \leq \xi \leq 2\pi : h = \langle h \rangle (1-b) \end{array} \right\}, \varepsilon_{\min} = 1 - b^2. \tag{2.37b}$$

A comparison of the two specified laws of variation of the THTC shows that the maximal effect of influence of a solid body on the average heat transfer for the step law is expressed much more strongly, than that for the cosine law (at identical relative amplitude of oscillations b). So, for example, for the value of $b = 0.9$ we have for the step function $h(\xi)$ about fivefold decrease in heat transfer rate, while for the cosine functions $h(\xi)$ this decrease is only double.

Asymptotical solution $\delta \to 0$. In this limiting case, a transfer of the external TBC from the heated surface $X = 0$ onto the heat transferring surface $X = \delta$ takes place. At our estimations, we shall base on that fact that in any point of a body the oscillation of temperatures $\hat{\vartheta}$ and heat fluxes $\hat{q} \sim k\hat{\vartheta}/\delta$ are limited. We can consider in the beginning the TBC of $\vartheta_0 = \text{const}\,(\theta \to 0)$ [(2.9a)]. Thus, it can be obtained from (2.25) that:

$$\varepsilon \to 1. \tag{2.38}$$

This effectively means (like in the case of $k \to \infty$ considered above) that the effect of thermal influence of a solid body is leveled off. For the TBC of $q_0 = \text{const}\,(\theta^{\bullet} \to 0)$, it follows from (2.31):

$$\varepsilon \to \varepsilon_{\min} = \left\langle \frac{1}{1+\psi} \right\rangle^{-1}. \tag{2.39}$$

The effect of the thermal influence of a solid body reaches here its maximum. Thus, the case of the negligible small wall thickness for the TBC $q_0 = \text{const}$ is equivalent to the case of a wall with a negligible small thermal conductivity.

References

1. Labuntsov DA, Zudin YB (1977) Peculiarities of the process of heat transfer from a surface of a plate to a flow with a spatiotemporal periodic variation of the heat transfer coefficient. Part 1. General analysis. Works of Moscow Power Engineering Institute. Issue 347: 84–92 (in Russian).
2. Labuntsov DA, Zudin YB (1977) Peculiarities of the process of heat transfer from a surface of a plate to a flow with a spatiotemporal periodic variation of the heat transfer coefficient. Part 2. Solution of characteristic problems. Works of Moscow Power Engineering Institute. Issue 347: 93–100 (in Russian).
3. Zudin YB (1980) Analysis of heat-transfer processes of periodic intensity. Dissertation. Moscow Power Engineering Institute (in Russian).
4. Labuntsov DA, Zudin YB (1984) Heat-Transfer Processes of Periodic Intensity. Energoatomizdat, Moscow (in Russian).
5. Stephenson G (1986) Partial Differential Equations for Scientists and Engineers. Longmann, London.
6. Zauderer E (1989) Partial Differential Equations of Applied Mathematics. Wiley, New York.
7. Sagan H (1989) Boundary and Eigenvalue Problems in Mathematical Physics. Dover Publications, New York.
8. Baehr HD, Stephan J (1998) Heat and Mass Transfer. Springer, Berlin Heidelberg New York.

3

Solution of Characteristic Problems

3.1 Construction of the General Solution

The main objective of the analysis presented in this chapter consists in finding solutions allowing calculating the factor of conjugation (FC), which definition is given by (2.13) [1–15]. After substitution of the equations for the temperature oscillations (2.9a), (2.9b) together with (2.11) for the THTC in the boundary condition (BC) (2.10), multiplication of the infinite Fourier series in the left-hand side of (2.10) and orthogonalization [16], it is possible in principle to determine both complex conjugate eigenvalues of this boundary problem A_n, A_n^* and the FC.

Let us write down (with the help of (2.9a) and (2.9b)) expressions for the fluctuation temperatures and temperature gradients at $X = \delta$

$$\theta = \sum_{n=1}^{\infty} [A_n \exp(in\,\xi) + A_n^* \exp(-in\,\xi)], \tag{3.1}$$

$$\theta^\bullet = \sum_{n=1}^{\infty} [B_n\, A_n \exp(in\,\xi) + B_n^*\, A_n^* \exp(-in\,\xi)]. \tag{3.2}$$

The following parameters are introduced herewith: $\theta = (\hat{\vartheta}_\delta)/\langle\vartheta_\delta\rangle$, dimensionless oscillations temperature;

$$\theta^\bullet = \frac{1}{\langle\vartheta_\delta\rangle} \left.\frac{\partial \hat{\vartheta}}{\partial x}\right|_{x=\bar\delta}$$

dimensionless gradient of the oscillating temperature (or dimensionless heat flux density); $B_n = F_n + i\Phi_n$, $B_n^* = F_n - i\Phi_n$, complex conjugate eigenfunctions of the boundary problem. The functions F_n, Φ_n describing real and imaginary parts of the eigenfunctions B_n, B_n^*, respectively, take the following form:

(a) For the TBC $\vartheta_0 = \text{const}$

$$F_n = \frac{r_n \sinh(2r_n\bar{\delta}) + s_n \sin(2s_n\bar{\delta})}{\cosh(2r_n\bar{\delta}) - \cos(2s_n\bar{\delta})}, \quad \Phi_n = \frac{s_n \sinh(2r_n\bar{\delta}) - r_n \sin(2s_n\bar{\delta})}{\cosh(2r_n\bar{\delta}) - \cos(2s_n\bar{\delta})}, \quad (3.3)$$

(b) for the TBC $q_0 = \text{const}$

$$F_n = \frac{r_n \sinh(2r_n\bar{\delta}) - s_n \sin(2s_n\bar{\delta})}{\cosh(2r_n\bar{\delta}) + \cos(2s_n\bar{\delta})}, \quad \Phi_n = \frac{s_n \sinh(2r_n\bar{\delta}) + r_n \sin(2s_n\bar{\delta})}{\cosh(2r_n\bar{\delta}) + \cos(2s_n\bar{\delta})}. \quad (3.4)$$

We shall refer below to F_n, Φ_n as to the *functions of thickness*. Their more detailed description is documented in Appendix B.

According to what we said above, let us present the THTC as

$$h = \langle h \rangle (1 + \psi). \quad (3.5)$$

A periodic part of the THTC can be found according to (2.21)

$$\psi = \sum_{n=1}^{\infty} [C_n \exp(in\xi) + C_n^* \exp(-in\xi)]. \quad (3.6)$$

Having substituted values $\theta, \theta^\bullet, \psi$ in the BC (2.29) and having executed multiplication of the infinite Fourier series, we obtain a general solution for the FC as

$$\varepsilon = 1 + \sum_{n=1}^{\infty} (C_n A_n + C_n^* A_n^*). \quad (3.7)$$

One can find eigenvalues A_n, A_n^* of the present boundary problem via equating the coefficients at corresponding exponential terms $\exp(\pm in\xi)$

$$\sum_{n=1}^{N}(C_n A_{N-n}) + \sum_{n=1}^{\infty}(C_n^* A_{N+n}) + \sum_{n=N+1}^{\infty}(C_n A_{n-N}) + C_N + \left(1 + \frac{B_N}{\langle \bar{h} \rangle}\right)A_N = 0, \quad (3.8)$$

$$\sum_{n=1}^{N}(C_n^* A_{N-n}^*) + \sum_{n=1}^{\infty}(C_n A_{N+n}^*) + \sum_{n=N+1}^{\infty}(C_n^* A_{n-N}^*) + C_N^* + \left(1 + \frac{B_N^*}{\langle \bar{h} \rangle}\right)A_N^* = 0, \quad (3.9)$$

$n = 1, 2, 3, \ldots, N = 1, 2, 3, \ldots$.

In principle, the system of (3.7)–(3.9) allows determining all eigenvalues of the boundary problem A_n, A_n^* (generally an infinite set), as well as the FC, which is actually the key value of the whole analysis. The sequence of the solution is as follows:

- According to the basic assumption of the method, the values C_n, C_n^* are to be considered preset
- From a solution of the infinite system of algebraic equations set by recurrent formulas (3.8), (3.9), one can obtain the values of A_n, A_n^*

– After substitution of the eigenvalues A_n, A_n^* into (3.7), the sought value of the FC can be determined

One should point out that, in fact, in spite of the linearity of the heat conduction equation (2.7), a use of the unsteady BC (3.5) transfers the investigated boundary problem in the nonlinear class. As it is known, such a problem has no exact analytical solution [17]. An analysis of (3.8) and (3.9) also shows that for any periodic function $\psi(\xi)$ it is not possible to obtain recurrent formulas for the values A_n, A_n^*. Therefore, strictly saying, the system of (3.7)–(3.9) represents not the solution itself, but only a construction of the general solution of the problem under investigation. An impossibility to derive an exact analytical solution of this problem in a general form results in the necessity of consecutive solutions of (3.7)–(3.9) for different characteristic functions $\psi(\xi)$. Let us consider in the beginning the simplest kind of oscillations of the heat transfer intensity described by a harmonic law.

3.2 Harmonic Law of Oscillations

A harmonic law of oscillations of the THTC (Fig. 3.1) can be set by the equation

$$\psi = b \cos \xi = \frac{b}{2}\left[\exp(i\xi) + \exp(-i\xi)\right]. \tag{3.10}$$

As it is known [18], heat transfer rate at turbulent fluid flow in a pipe can be estimated from the equation $Nu = 0.023 Re^{0.8} Pr^{0.4}$, where $Nu = hD/k_\mathrm{f}$ is the Nusselt number, $Re = uD/\nu_\mathrm{f}$ is the Reynolds number, $Pr = \nu_\mathrm{f}/\alpha_\mathrm{f}$ is the Prandtl number, D is the pipe's diameter. It follows from here that in this case the THTC depends on the flow velocity as $h \sim u^{0.8}$. If the basic flow is subjected to external harmonic oscillations, which frequency is much less than frequency of the turbulent vortices generation [19], the structure of turbulence practically does not change, and the heat transfer intensity will also undergo oscillations close by their form to a harmonic law.

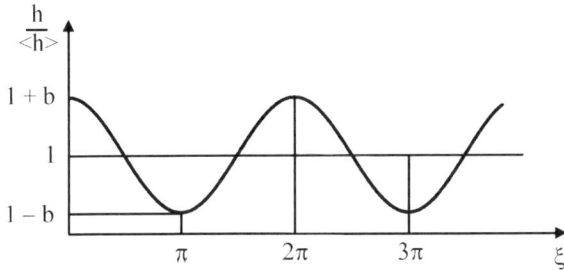

Fig. 3.1. Harmonic law of pulsations of the THTC

3 Solution of Characteristic Problems

General solution for the harmonic oscillation. A substitution of (3.1), (3.2), and (3.10) into BC (2.29) gives

$$\left.\begin{aligned}\varepsilon &= 1 + \tfrac{b}{2}(A_1 + A_1^*), \\ \left(1 + \tfrac{B_1}{\langle \bar h \rangle}\right) A_1 + \tfrac{b}{2}(1 + A_2) &= 0, \\ \left(1 + \tfrac{B_1^*}{\langle \bar h \rangle}\right) A_1^* + \tfrac{b}{2}(1 + A_2^*) &= 0, \\ &\vdots \\ \left(1 + \tfrac{B_n}{\langle \bar h \rangle}\right) A_n + \tfrac{b}{2}(A_{n-1} + A_{n+1}) &= 0, \\ \left(1 + \tfrac{B_n^*}{\langle \bar h \rangle}\right) A_n^* + \tfrac{b}{2}(A_{n-1}^* + A_{n+1}^*) &= 0.\end{aligned}\right\} \quad (3.11)$$

A solution of this infinite system of algebraic equations can be obtained using a method of induction and can be finally written as a sum of two infinite complex conjugate chain fractions [20][1]

$$\left.\begin{aligned}\varepsilon &= 1 + \tfrac{b}{2}(s + s^*), \\ s = c_0 - \cfrac{1}{c_1 - \cfrac{1}{c_2 - \cfrac{1}{c_3 - \dots}}}, \quad s^* = c_0^* - \cfrac{1}{c_1^* - \cfrac{1}{c_2^* - \cfrac{1}{c_3^* - \dots}}}.\end{aligned}\right\} \quad (3.12)$$

Here $c_0 = c_0^* = 0$,

$$c_n = \frac{2}{b}\left(1 + \frac{B_n}{\langle \bar h \rangle}\right), \quad c_n^* = \frac{2}{b}\left(1 + \frac{B_n^*}{\langle \bar h \rangle}\right),$$

with $n = 1, 2, 3, \dots$ are recurrent formulas for n-term of the chain fraction. Let us write also another form of notation of the chain fractions, i.e., through n-partial sums s_n, s_n^* [20]

$$\left.\begin{aligned}s = s_0 &= c_0 - \tfrac{1}{s_1}, \quad s^* = s_0^* = c_0^* - \tfrac{1}{s_1^*}, \\ s_1 &= c_1 - \tfrac{1}{s_2}, \quad s_1^* = c_1^* - \tfrac{1}{s_2^*}, \\ s_2 &= c_2 - \tfrac{1}{s_3}, \quad s_2^* = c_2^* - \tfrac{1}{s_3^*}, \\ &\vdots \\ s_n &= c_n - \tfrac{1}{s_{n+1}}, \quad s_n^* = c_n^* - \tfrac{1}{s_{n+1}^*}.\end{aligned}\right\} \quad (3.13)$$

Asymptotical cases for expressions (3.12) are considered below.
Asymptotical solution at $\langle \bar h \rangle \to 0$. Here n-terms of a chain fraction will be transformed to the notation of

---

[1] The theory of positive chain fractions is based on three fundamental theorems that are proved in [20] using the method of mathematical induction. A generalization of one of these fundamental theorems for the case of the chain fraction with an arbitrary sign is given in Appendix C.

$$c_n = \frac{2}{b} \frac{B_n}{\langle \bar{h} \rangle}, \quad c_n^* = \frac{2}{b} \frac{B_n^*}{\langle \bar{h} \rangle}.$$

Having multiplied and divided the values s_n, s_n^* by

$$\frac{b}{2} \frac{B_n}{\langle \bar{h} \rangle}, \quad \frac{b}{2} \frac{B_n^*}{\langle \bar{h} \rangle},$$

one can rewrite the system of (3.12) as

$$\left.\begin{aligned}
\varepsilon &= 1 - \frac{b^2}{4} \langle \bar{h} \rangle \left(\frac{1}{B_1 S_1} + \frac{1}{B_1^* S_1^*} \right), \\
S_1 &= 1 + \frac{\langle \bar{h} \rangle}{B_1} - \frac{b^2}{4} \frac{\langle \bar{h} \rangle^2}{B_1 B_2 S_2}, \quad S_1^* = 1 + \frac{\langle \bar{h} \rangle}{B_1^*} - \frac{b^2}{4} \frac{\langle \bar{h} \rangle^2}{B_1^* B_2^* S_2^*}, \\
S_2 &= 1 + \frac{\langle \bar{h} \rangle}{B_2} - \frac{b^2}{4} \frac{\langle \bar{h} \rangle^2}{B_2 B_3 S_3}, \quad S_2^* = 1 + \frac{\langle \bar{h} \rangle}{B_2^*} - \frac{b^2}{4} \frac{\langle \bar{h} \rangle^2}{B_2^* B_3^* S_3^*}, \\
&\vdots \\
S_n &= 1 + \frac{\langle \bar{h} \rangle}{B_n} - \frac{b^2}{4} \frac{\langle \bar{h} \rangle^2}{B_n B_{n+1} S_{n+1}}, \quad S_n^* = 1 + \frac{\langle \bar{h} \rangle}{B_n^*} - \frac{b^2}{4} \frac{\langle \bar{h} \rangle^2}{B_n^* B_{n+1}^* S_{n+1}^*}, \\
& n = 1, 2, 3, \ldots.
\end{aligned}\right\} \quad (3.14)$$

Let us find a derivative $\varepsilon' = d\varepsilon/d\langle \bar{h} \rangle$:

$$\varepsilon' = -\frac{b^2}{4} \left(\frac{1}{B_1 S_1} + \frac{1}{B_1^* S_1^*} \right) + \frac{b^2}{4} \left(\frac{S_1'}{B_1 S_1^2} + \frac{S_1^{*'}}{B_1^* S_1^{*2}} \right) \langle \bar{h} \rangle. \quad (3.15)$$

Having consecutively differentiated values S_n, S_n^* with respect to $\langle \bar{h} \rangle$ one can find recurrent formulas for the derivatives $S_1', S_1^{*'}$, which are too cumbersome and therefore not given here. Assuming that $\langle \bar{h} \rangle = 0$ in the obtained relations, one can find: $S_n(0) = S_n^*(0) = 1, S_n'(0) = 1/B_n, S_n^{*'}(0) = 1/B_n^*$. Proceeding further, one can obtain the following value of a derivative of the FC with respect to $\langle \bar{h} \rangle$:

$$\varepsilon'(0) = -\frac{b^2}{4} \left(\frac{1}{B_1} + \frac{1}{B_1^*} \right). \quad (3.16)$$

It follows from here in view of the equality $B_n = F_n + i\Phi_n, B_n^* = F_n - i\Phi_n$ that the required asymptotical form of solution (3.12) at $\langle \bar{h} \rangle \to 0$ can be written as

$$\varepsilon = 1 - \frac{b^2}{2} \frac{F_1}{F_1^2 + \Phi_1^2} \langle \bar{h} \rangle. \quad (3.17)$$

This equation represents the first two terms of a Taylor series expansion of the function $\varepsilon(\langle \bar{h} \rangle)$ around a point $\langle \bar{h} \rangle \to 0$. An important special case of this problem represents heat transfer on a surface of a semi-infinite body ($\delta \to \infty$). Here we have $F_n = n, \Phi_n = 0$ for a spatial problem and $F_n = \Phi_n = \sqrt{n/2}$ for a time-dependent problem, respectively. For these particular cases, the asymptotical solution takes the following form: a spatial problem

$$\varepsilon = 1 - \frac{b^2}{2} \langle \bar{h} \rangle, \quad \langle \bar{h} \rangle = \frac{\langle h \rangle Z_0}{k}, \quad (3.17a)$$

a time-dependent problem

$$\varepsilon = 1 - \frac{b^2}{2\sqrt{2}} \langle \tilde{h} \rangle, \quad \langle \tilde{h} \rangle = \frac{\langle h \rangle \sqrt{\alpha \tau_0}}{k}. \tag{3.17b}$$

Asymptotical solution at $\langle \bar{h} \rangle \to \infty$. Let us construct a Taylor series expansion of the function $\varepsilon(\langle \bar{h} \rangle)$ about a point $\beta = \langle \bar{h} \rangle^{-1} \to 0$. One can rewrite the system of (3.12) as

$$\left.\begin{array}{l} \varepsilon = 1 - \frac{b^2}{4}\left(\frac{1}{S_1} + \frac{1}{S_1^*}\right), \\ S_1 = 1 + B_1\beta - \frac{b^2}{4}\frac{1}{S_2}, \quad S_1^* = 1 + B_1^*\beta - \frac{b^2}{4}\frac{1}{S_2^*}, \\ S_2 = 1 + B_2\beta - \frac{b^2}{4}\frac{1}{S_3}, \quad S_2^* = 1 + B_2^*\beta - \frac{b^2}{4}\frac{1}{S_3^*}, \\ \vdots \\ S_n = 1 + \frac{\langle \bar{h} \rangle}{B_n} - \frac{b^2}{4}\frac{\langle \bar{h} \rangle^2}{B_n B_{n+1} S_{n+1}}, \quad S_n^* = 1 + \frac{\langle \bar{h} \rangle}{B_n^*} - \frac{b^2}{4}\frac{\langle \bar{h} \rangle^2}{B_n^* B_{n+1}^* S_{n+1}^*}, \\ n = 1, 2, 3, \ldots . \end{array}\right\} \tag{3.18}$$

Let us find a derivative of the FC with respect to a small parameter $\varepsilon' = d\varepsilon/d\beta$:

$$\varepsilon' = \frac{b^2}{4}\left(\frac{S_1'}{S_1^2} + \frac{S_1^{*'}}{S_1^{*2}}\right). \tag{3.19}$$

Consecutively differentiating all values S_n, S_n^* with respect to β, one can find recurrent formulas for the derivatives $S_1', S_1^{*'}$

$$S_n' = B_n + \frac{b^2}{4}\frac{S_{n+1}'}{S_{n+1}^2}, \quad S_n^{*'} = B_n^* + \frac{b^2}{4}\frac{S_{n+1}^{*'}}{S_{n+1}^{*2}}. \tag{3.20}$$

Assuming $\beta = 0$ in these expressions, one can find: $S_n = 1 - b^2/(4S_n)$, $S_n^* = 1 - b^2/(4S_n^*)$. Then one can find the values S_n, S_n^* from the solution of the respective quadratic equations: $S_n = S_n^* = (1 + \sqrt{1-b^2})/2$. As a result of these derivations, one can obtain the first two terms of a Taylor series expansion of the function $\varepsilon(\langle \bar{h} \rangle)$ about a point $\langle \bar{h} \rangle \to \infty$

$$\varepsilon = \sqrt{1-b^2} + 2\sum_{n=1}^{\infty}\left[\left(\frac{b}{1+\sqrt{1-b^2}}\right)^{2n} F_n\right]\langle \bar{h} \rangle^{-1}. \tag{3.21}$$

The use of (3.3) and (3.4) for the function F_n results in the fact that an infinite series in this relationship is always diverging. This means further that the operations of differentiation of the chain fractions executed above do not satisfy the convergence conditions [20]. Thus, the solution (3.12) has no analytical asymptotical form at $\langle \bar{h} \rangle \to \infty$.

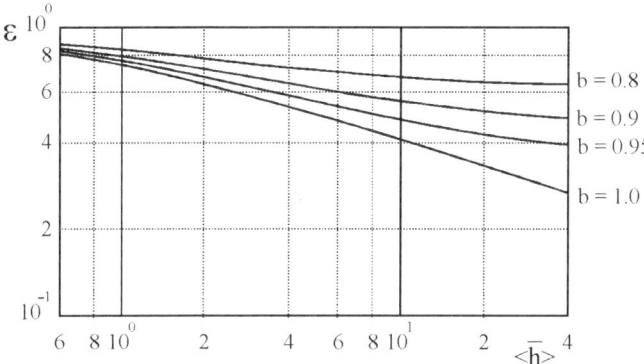

Fig. 3.2. Harmonic pulsations. Spatial problem for a semi-infinite body. Values of the factor of conjugation

Asymptotical solution at $\delta \to 0$ *for the TBC* $\vartheta_0 = $ const. In this asymptotical solution, the reasoning similar to previously mentioned one results in the following series expansion for the FC

$$\varepsilon = 1 - \frac{b^2}{2} \langle \bar{h} \rangle \bar{\delta}, \qquad (3.22)$$

or, in a more obvious form

$$\varepsilon = 1 - \frac{b^2}{2} \frac{\langle h \rangle \delta}{k}. \qquad (3.23)$$

Asymptotical solution at $b \to 0$. This asymptotical solution describes negligibly small oscillations of the THTC ($\psi \to 0$). In this case, the thermal influence of a solid body degenerates, and that is physically obvious. Therefore, the EHTC becomes equal to the ATHTC, so that $\varepsilon = 1$.

Asymptotical solution at $b \to 1$. For a case of the maximal amplitude of oscillations of the THTC $\psi = \cos(\xi)$, the thermal influence of a body reaches its maximum. It is important to note, however, that for the considered harmonic law of variation of the THTC at $b = 1$, solution (3.12) does not exhibit any particular property. Results of computations of the FC for the harmonic law of oscillations of the THTC are documented in Tables 3.1 and 3.2 and in Figs. 3.2 and 3.3. The performed analysis confirms the conclusions made above at the estimation of the construction of the general solution of (3.7)–(3.9): even for the simplest type of oscillations, recurrent formulas for A_n, A_n^* can be notated as infinite complex conjugate chain fractions.

3.3 Inverse Harmonic Law of Oscillations

Let us consider now an inverse harmonic law of oscillations of the THTC (Fig. 3.4).

Table 3.1. Harmonic pulsations

$\langle \bar{h} \rangle$	b					
	1.0	0.95	0.90	0.80	0.70	0.60
0.1	0.9545	0.9589	0.9632	0.9709	0.9770	0.9836
0.2	0.9164	0.9245	0.9323	0.9465	0.9591	0.9700
0.3	0.8837	0.8952	0.9060	0.9258	0.9432	0.9584
0.4	0.8554	0.8697	0.8832	0.9079	0.9236	0.9484
0.5	0.8305	0.8473	0.8631	0.8922	0.9177	0.9396
0.6	0.8083	0.8274	0.8454	0.8783	0.9071	0.9320
0.7	0.7884	0.8096	0.8295	0.8659	0.8978	0.9252
0.8	0.7704	0.7934	0.8152	0.8548	0.8894	0.9191
0.9	0.7540	0.7788	0.8022	0.8447	0.8818	0.9136
1.0	0.7389	0.7654	0.7903	0.8356	0.8749	0.9086
2.0	0.6346	0.6734	0.7095	0.7742	0.8295	0.8762
3.0	0.5728	0.6200	0.6635	0.7403	0.8051	0.8591
4.0	0.5302	0.5839	0.6328	0.7183	0.7896	0.8485
5.0	0.4982	0.5571	0.6105	0.7027	0.7787	0.8411
6.0	0.4729	0.5364	0.5933	0.6910	0.7707	0.8358
7.0	0.4522	0.5196	0.5796	0.6818	0.7645	0.8316
8.0	0.4348	0.5057	0.5684	0.6743	0.7596	0.8284
9.0	0.4197	0.4939	0.5590	0.6682	0.7555	0.8257
10	0.4066	0.4837	0.5510	0.6630	0.7522	0.8236
20	0.3285	0.4260	0.5071	0.6363	0.7352	0.8128
30	0.2892	0.3995	0.4883	0.6257	0.7288	0.8088
40	0.2638	0.3838	0.4776	0.6198	0.7253	0.8067
50	0.2457	0.3732	0.4705	0.6162	0.7232	0.8054
60	0.2315	0.3654	0.4656	0.6137	0.7218	0.8046
70	0.2203	0.3595	0.4619	0.6119	0.7207	0.8039
80	0.2111	0.3549	0.4590	0.6105	0.7200	0.8034
90	0.2030	0.3512	0.4568	0.6094	0.7193	0.8031
100	0.1962	0.3490	0.4549	0.6085	0.7188	0.8082
200	0.1566	0.3323	0.4461	0.6044	0.7165	0.8014
∞	0	0.3122	0.4359	0.6000	0.7141	0.8000

Spatial problem for a semi-infinite body. Values of the factor of conjugation

$$\psi = \frac{\sqrt{1-b^2}}{1+b\cos\xi} = \frac{\sqrt{1-b^2}}{1+(b/2)\left[\exp(i\xi)+\exp(-i\xi)\right]}. \quad (3.24)$$

As it was mentioned above, at a laminar regime of flow of a liquid film on a heated surface, the THTC can be rather precisely described by the dependence $h = k_{\mathrm{f}}/\delta_{\mathrm{f}}$. In this case, harmonic oscillations of a film thickness can be observed already at small Reynolds numbers: $\delta_{\mathrm{f}} = \langle \delta_{\mathrm{f}} \rangle [1 + b\cos(\xi)]$. The phase speed of the wave propagation has an order of magnitude of an average velocity of a liquid in the film: $u \sim Z_0/\tau_0$. It can be concluded from here that indeed the oscillations of the THTC at a wave flow of a film with a good degree of accuracy can be described by an inverse harmonic (inverse cosine) law.

General solution for the inverse harmonic oscillations. A substitution of (3.1), (3.2), and (3.24) into BC (2.30) gives

3.3 Inverse Harmonic Law of Oscillations

Table 3.2. Harmonic pulsations

$\langle \tilde{h} \rangle$	b					
	1.0	0.95	0.90	0.80	0.70	0.60
0.1	0.9650	0.9684	0.9716	0.9776	0.9828	0.9874
0.2	0.9316	0.9383	0.9446	0.9562	0.9664	0.9753
0.3	0.9007	0.9103	0.9195	0.9363	0.9512	0.9641
0.4	0.8723	0.8847	0.8964	0.9181	0.8373	0.9539
0.5	0.8463	0.8612	0.8754	0.9015	0.9246	0.9445
0.6	0.8226	0.8399	0.8563	0.8864	0.9130	0.9361
0.7	0.8010	0.8204	0.8388	0.8726	0.9025	0.9284
0.8	0.7811	0.8025	0.8228	0.8601	0.8930	0.9214
0.9	0.7628	0.7861	0.8082	0.8487	0.8843	0.9151
1.0	0.7460	0.7710	0.7947	0.8382	0.8764	0.9094
2.0	0.6277	0.6661	0.7022	0.7675	0.8240	0.8720
3.0	0.5577	0.6055	0.6500	0.7293	0.7967	0.8531
4.0	0.5097	0.5649	0.6158	0.7053	0.7800	0.8418
5.0	0.4741	0.5354	0.5914	0.6887	0.7689	0.8345
6.0	0.4460	0.5127	0.5730	0.6766	0.7609	0.8283
7.0	0.4231	0.4946	0.5586	0.6674	0.7549	0.8254
8.0	0.4041	0.4797	0.5470	0.6601	0.7503	0.8224
9.0	0.3878	0.4671	0.5374	0.6542	0.7466	0.8200
10	0.3736	0.4566	0.6294	0.6494	0.7435	0.8181
20	0.2904	0.3990	0.4880	0.6259	0.7292	0.8092
30	0.2495	0.3747	0.4719	0.6175	0.7243	0.8062
40	0.2235	0.3610	0.4634	0.6132	0.7218	0.8046
50	0.2053	0.3522	0.4582	0.6106	0.7203	0.8037
60	0.1914	0.3462	0.4546	0.6089	0.7192	0.8031
70	0.1803	0.3417	0.4521	0.6076	0.7185	0.8027
80	0.1712	0.3383	0.4501	0.6067	0.7180	0.8023
90	0.1635	0.3356	0.4486	0.6059	0.7176	0.8021
100	0.1570	0.3334	0.4472	0.6054	0.7172	0.8019
200	0.1198	0.3235	0.4417	0.6027	0.7157	0.8009
∞	0	0.3122	0.4359	0.600	0.7141	0.8000

Time-dependent problem for a semi-infinite body. Values of the factor of conjugation

$$\left.\begin{array}{r}
\varepsilon = \sqrt{1-b^2} + \frac{b}{2}\frac{B_1 A_1 + B_1^* A_1^*}{\langle \tilde{h} \rangle}, \\
\frac{b}{2}\varepsilon = \left(\sqrt{1-b^2} + \frac{B_1}{\langle \tilde{h} \rangle}\right) A_1 + \frac{b}{2}\frac{B_2 A_2}{\langle \tilde{h} \rangle}, \\
\frac{b}{2}\varepsilon = \left(\sqrt{1-b^2} + \frac{B_1^*}{\langle \tilde{h} \rangle}\right) A_1^* + \frac{b}{2}\frac{B_2^* A_2^*}{\langle \tilde{h} \rangle}, \\
\left(1 + \frac{B_1}{\langle \tilde{h} \rangle}\right) A_1 + \frac{b}{2}(1 + A_2) = 0, \\
\left(1 + \frac{B_1^*}{\langle \tilde{h} \rangle}\right) A_1^* + \frac{b}{2}(1 + A_2^*) = 0, \\
\vdots \\
\left(1 + \frac{B_n}{\langle \tilde{h} \rangle}\right) A_n + \frac{b}{2}(A_{n-1} + A_{n+1}) = 0, \\
\left(1 + \frac{B_n^*}{\langle \tilde{h} \rangle}\right) A_n^* + \frac{b}{2}(A_{n-1}^* + A_{n+1}^*) = 0, \\
n = 2, 3, 4, \ldots.
\end{array}\right\} \quad (3.25)$$

46   3 Solution of Characteristic Problems

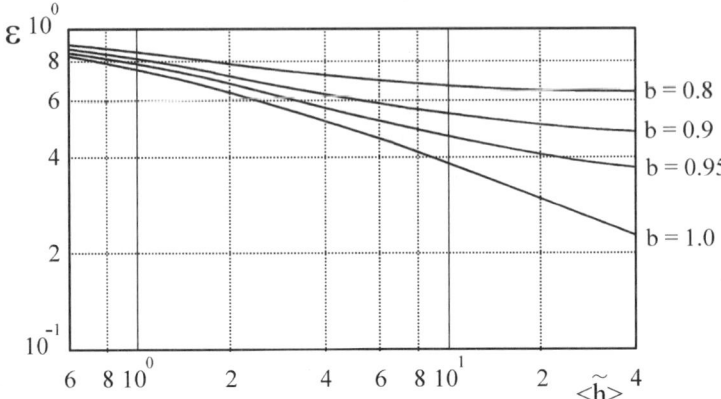

Fig. 3.3. Harmonic pulsations. Time-dependent problem for a semi-infinite body. Values of the factor of conjugation

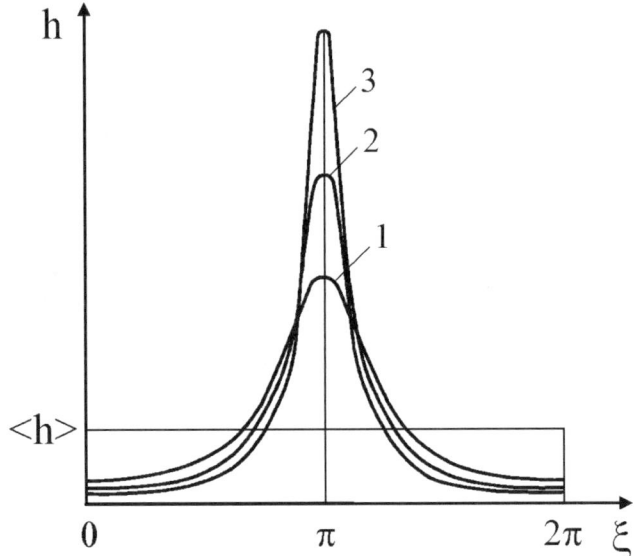

Fig. 3.4. Inverse harmonic law of pulsations of the THTC: $1-b=0.5, 2-b=0.9$, $3-b=0.95$

Like in the case of a harmonic law, the solution of the written above infinite system of algebraic equations can be derived using the method of induction and can be ultimately written as a sum of two infinite complex conjugate chain fractions [20]:

$$\left.\begin{array}{l}\varepsilon = \frac{\sqrt{1-b^2}}{1+(b/2)(S+S^*)}, \\ S = c_0 - \cfrac{1}{c_1 - \cfrac{1}{c_2 - \cfrac{1}{c_3 - \ldots}}}, \quad S^* = c_0^* - \cfrac{1}{c_1^* - \cfrac{1}{c_2^* - \cfrac{1}{c_3^* - \ldots}}}.\end{array}\right\} \quad (3.26)$$

3.3 Inverse Harmonic Law of Oscillations

Table 3.3. Inverse harmonic pulsations. Spatial problem for a semi-infinite body. Values of the factor of conjugation

$\langle \bar{h} \rangle$	b				
	0.95	0.90	0.80	0.70	0.60
0.1	0.8674	0.9116	0.9488	0.9676	0.9788
0.2	0.7894	0.8520	0.9107	0.9421	0.9619
0.3	0.7316	0.8063	0.8804	0.9214	0.9480
0.4	0.6836	0.7701	0.8556	0.9039	0.9363
0.5	0.6490	0.7406	0.8350	0.8896	0.9264
0.6	0.6206	0.7149	0.8176	0.8772	0.9178
0.7	0.5969	0.6945	0.8026	0.8666	0.9103
0.8	0.5768	0.6770	0.7897	0.8572	0.9038
0.9	0.5594	0.6619	0.7783	0.8490	0.8980
1.0	0.5443	0.6486	0.7680	0.8417	0.8928
2.0	0.4581	0.5708	0.7083	0.7971	0.8608
3.0	0.4200	0.5357	0.6803	0.7758	0.8453
4.0	0.3980	0.5153	0.6639	0.7633	0.8361
5.0	0.3837	0.5020	0.6531	0.7550	0.8301
6.0	0.3736	0.4925	0.6455	0.7491	0.8257
7.0	0.3660	0.4854	0.6398	0.7447	0.8225
8.0	0.3602	0.4800	0.6353	0.7413	0.8200
9.0	0.3555	0.4756	0.6318	0.7386	0.8180
10	0.3516	0.4720	0.6289	0.7364	0.8164
20	0.3332	0.4549	0.6152	0.7258	0.8086
30	0.3265	0.4488	0.6103	0.7220	0.8058
40	0.3231	0.4457	0.6078	0.7201	0.8044
50	0.3210	0.4438	0.6063	0.7189	0.8035
60	0.3196	0.4425	0.6052	0.7182	0.8030
70	0.3158	0.4416	0.6045	0.7176	0.8025
80	0.3178	0.4409	0.6039	0.7172	0.8022
90	0.3172	0.4403	0.6035	0.7168	0.8020
100	0.3167	0.4399	0.6032	0.7166	0.8018
200	0.3145	0.4379	0.6016	0.7154	0.8009
∞	0.3122	0.4359	0.6000	0.7141	0.8000

Here

$$c_n = \frac{2}{b}\left(1 + \frac{B_n}{\langle \bar{h} \rangle}\right), \quad c_n^* = \frac{2}{b}\left(1 + \frac{B_n^*}{\langle \bar{h} \rangle}\right),$$

with $n = 1, 2, 3, \ldots$ are recurrent formulas for an n-term of the chain fraction. Another form of notating the chain fractions [20] (via n-partial sums S_n, S_n^*) is similar to the case of the harmonic law, and therefore it is not written here. Results of computations of the FC for the inverse harmonic law of oscillations of the THTC are shown in Tables 3.3 and 3.4 and in Figs. 3.5 and 3.6. An analysis of asymptotical forms of the solution (3.26) and effects of different determining parameters in them is given below.

Asymptotical solution at $\langle \bar{h} \rangle \to 0$. Let us rewrite the system of equations (3.26) as

Table 3.4. Inverse harmonic pulsations

$\langle \tilde{h} \rangle$	\multicolumn{5}{c}{b}				
	0.95	0.90	0.80	0.70	0.60
0.1	0.8809	0.9240	0.9578	0.9740	0.9832
0.2	0.8044	0.8670	0.9224	0.9506	0.9679
0.3	0.7745	0.8202	0.8918	0.9300	0.9541
0.4	0.6953	0.7816	0.8656	0.9118	0.9419
0.5	0.6578	0.7490	0.8432	0.8961	0.9310
0.6	0.6268	0.7224	0.8238	0.8823	0.9215
0.7	0.6009	0.6998	0.8071	0.8702	0.9130
0.8	0.5789	0.6804	0.7925	0.8596	0.9055
0.9	0.5600	0.6635	0.7796	0.8502	0.8988
1.0	0.5437	0.6487	0.7684	0.8418	0.8928
2.0	0.4521	0.5644	0.7018	0.7917	0.8566
3.0	0.4126	0.5272	0.6722	0.7691	0.8401
4.0	0.3902	0.5065	0.6556	0.7564	0.8309
5.0	0.3758	0.4932	0.6451	0.7484	0.8250
6.0	0.3659	0.4641	0.6378	0.7429	0.8210
7.0	0.3586	0.4774	0.6326	0.7389	0.8181
8.0	0.3530	0.4724	0.6286	0.7359	0.8159
9.0	0.3486	0.4684	0.6254	0.7335	0.8142
10	0.3450	0.4652	0.6229	0.7316	0.8128
20	0.3286	0.4505	0.6114	0.7229	0.8064
30	0.3231	0.4456	0.6076	0.7200	0.8043
40	0.3204	0.4431	0.6057	0.7185	0.8032
50	0.3187	0.4417	0.6046	0.7176	0.8026
60	0.3176	0.4407	0.6038	0.7170	0.8021
70	0.3169	0.4400	0.6032	0.7166	0.8018
80	0.3163	0.4395	0.6028	0.7163	0.8016
90	0.3158	0.4391	0.6025	0.7161	0.8014
100	0.3155	0.4388	0.6023	0.7159	0.8013
200	0.3138	0.4373	0.6011	0.7150	0.8006
∞	0.3122	0.4359	0.6000	0.7141	0.8000

Time-dependent problem for a semi-infinite body. Values of the factor of conjugation

$$\left.\begin{aligned}
\varepsilon &= \frac{\sqrt{1-b^2}}{1-(b^2/4)\left(1/S_1+1/S_1^*\right)}, \\
S_1 &= 1 + \frac{\sqrt{1-b^2}\langle \bar{h}\rangle}{B_1} - \frac{b^2}{4S_2}, \quad S_1^* = 1 + \frac{\sqrt{1-b^2}\langle \bar{h}\rangle}{B_1^*} - \frac{b^2}{4S_2^*}, \\
S_2 &= 1 + \frac{\sqrt{1-b^2}\langle \bar{h}\rangle}{B_2} - \frac{b^2}{4S_3}, \quad S_2^* = 1 + \frac{\sqrt{1-b^2}\langle \bar{h}\rangle}{B_2^*} - \frac{b^2}{4S_3^*}, \\
&\qquad\qquad \vdots \\
S_n &= 1 + \frac{\sqrt{1-b^2}\langle \bar{h}\rangle}{B_n} - \frac{b^2}{4S_{n+1}}, \quad S_n^* = 1 + \frac{\sqrt{1-b^2}\langle \bar{h}\rangle}{B_n^*} - \frac{b^2}{4S_{n+1}^*}, \\
&\qquad n = 1, 2, 3, \dots.
\end{aligned}\right\} \quad (3.27)$$

A procedure similar to that used above for the harmonic law, gives the following series expansion

$$\varepsilon = 1 - 2\sum_{n=1}^{\infty}\left[\left(\frac{b}{1+\sqrt{1-b^2}}\right)^{2n} \frac{F_n}{F_n^2+\Phi_n^2}\right]\langle \bar{h}\rangle. \quad (3.28)$$

3.3 Inverse Harmonic Law of Oscillations   49

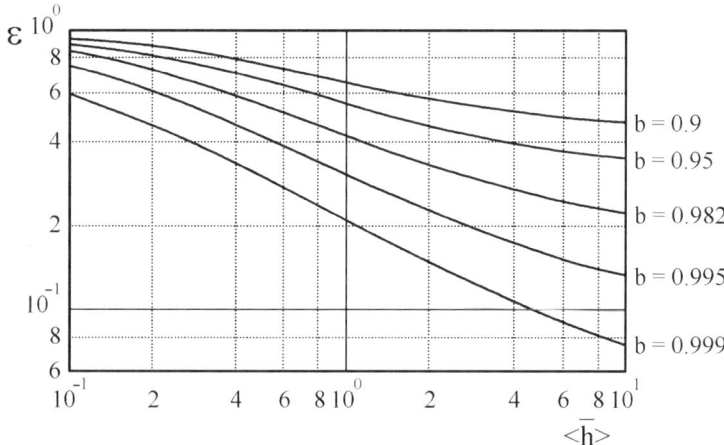

Fig. 3.5. Inverse harmonic pulsations. Spatial problem for a semi-infinite body. Values of the factor of conjugation

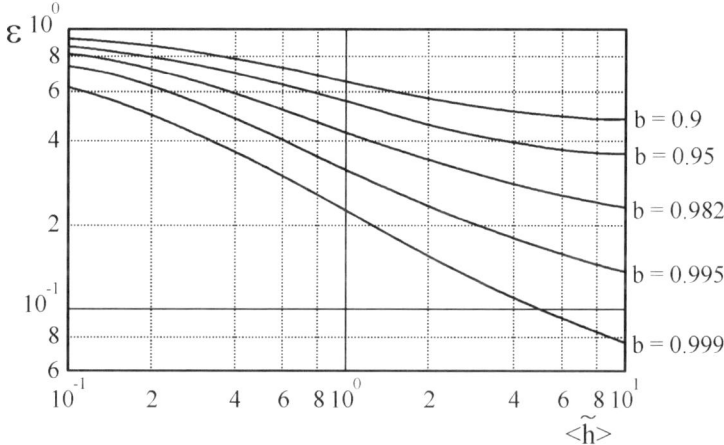

Fig. 3.6. Inverse harmonic pulsations. Time-dependent problem for a semi-infinite body. Values of the factor of conjugation

For a case of semi-infinite bodies $\delta \to \infty$ one can obtain: for a spatial problem

$$\varepsilon = 1 - 2\sum_{n=1}^{\infty}\left[\left(\frac{b}{1+\sqrt{1-b^2}}\right)^{2n}\frac{1}{n}\right]\langle \bar{h}\rangle, \qquad (3.28a)$$

for a time-dependent problem

$$\varepsilon = 1 - \sqrt{2}\sum_{n=1}^{\infty}\left[\left(\frac{b}{1+\sqrt{1-b^2}}\right)^{2n}\frac{1}{\sqrt{n}}\right]\langle \tilde{h}\rangle. \qquad (3.28b)$$

Let us show that infinite series in the right-hand sides of these relations are converging. Since

$$\left(\frac{b}{1+\sqrt{1-b^2}}\right)^{2n} \geq \left(\frac{b}{1+\sqrt{1-b^2}}\right)^{2n}\frac{1}{n}, \qquad (3.29)$$

$$\left(\frac{b}{1+\sqrt{1-b^2}}\right)^{2n} \geq \left(\frac{b}{1+\sqrt{1-b^2}}\right)^{2n}\frac{1}{\sqrt{n}}, \qquad (3.30)$$

then both power series under investigation will be limited from above by the following series

$$\sum_{n=1}^{\infty}\left(\frac{b}{1+\sqrt{1-b^2}}\right)^{2n} = \frac{1-\sqrt{1-b^2}}{2\sqrt{1-b^2}}, \qquad (3.31)$$

which represents a sum of an indefinitely decreasing geometrical progression. From this fact, convergence of these series follows at $b \leq 1$. At $b \leq 1$, the series are diverging

$$\sum_{n=1}^{\infty}\frac{1}{n} = \infty, \quad \sum_{n=1}^{\infty}\frac{1}{\sqrt{n}} = \infty. \qquad (3.32)$$

Asymptotical solution at $\langle \bar{h} \rangle \to \infty$. The reasoning similar to the aforementioned one gives

$$\varepsilon = \sqrt{1-b^2} + \frac{b^2}{2}\frac{F_1}{\langle \bar{h} \rangle}. \qquad (3.33)$$

For the case of the semi-infinite body ($\delta \to \infty$), one can obtain: for a spatial problem

$$\varepsilon = \sqrt{1-b^2} + \frac{b^2}{2}\frac{1}{\langle \bar{h} \rangle}, \qquad (3.33a)$$

for a time-dependent problem

$$\varepsilon = \sqrt{1-b^2} + \frac{b^2}{2\sqrt{2}}\frac{1}{\langle \tilde{h} \rangle}. \qquad (3.33b)$$

Asymptotical solution at $\delta \to 0$ for the TBC $\vartheta_0 = \text{const}$ is given by a relation

$$\varepsilon = 1 - \frac{1-\sqrt{1-b^2}}{2\sqrt{1-b^2}}\langle \bar{h} \rangle \bar{\delta}, \qquad (3.34)$$

or, in a more obvious form,

$$\varepsilon = 1 - \frac{1-\sqrt{1-b^2}}{2\sqrt{1-b^2}}\frac{\langle h \rangle \delta}{k}. \qquad (3.35)$$

3.3 Inverse Harmonic Law of Oscillations

Asymptotical solution at $\delta \to 0$ *for the TBC* $q_0 = \text{const}$ *looks like: for a spatial problem*

$$\varepsilon = \sqrt{1-b^2} + \frac{b^2}{2}\frac{\bar{\delta}}{\langle \bar{h} \rangle}, \qquad (3.35a)$$

for a time-dependent problem

$$\varepsilon = \sqrt{1-b^2} + \frac{b^2}{2}\frac{\tilde{\delta}}{\langle \tilde{h} \rangle}. \qquad (3.35b)$$

Asymptotical solution at $b \to 0$ *for the TBC* $\vartheta_0 = \text{const}$. In this case, oscillations of the heat transfer intensity are negligibly small: $\psi \to 0$. As a result, the thermal influence of a solid body vanishes: $\varepsilon \to 1$.

Asymptotical solution at $b \to 1$. The form of the function $\psi(\xi)$ with increasing b becomes more and more asymmetric (Fig. 3.7). Transition to the limiting case of $b = 1$ is accompanied with a qualitative transformation of the nature of oscillations of the THTC, which can be described in this case by the Kroeneker delta-function [21]

$$\left.\begin{array}{l} \xi = \pi : h = \infty \\ \xi \neq \pi : h = 0 \end{array}\right\}. \qquad (3.36)$$

Let us linearize n-partial sums S_n, S_n^* by means of a small parameter $\gamma = 2\sqrt{(1-b)/(1+b)}$:

$$\left.\begin{array}{ll} S_1 = 1 + \frac{\gamma}{B_1} - \frac{1}{4S_2}, & S_1^* = 1 + \frac{\gamma}{B_1^*} - \frac{1}{4S_2^*}, \\ S_2 = 1 + \frac{\gamma}{B_2} - \frac{1}{4S_3}, & S_2^* = 1 + \frac{\gamma}{B_2^*} - \frac{1}{4S_3^*}, \\ \quad \vdots & \\ S_n = 1 + \frac{\gamma}{B_n} - \frac{1}{4S_{n+1}}, & S_n^* = 1 + \frac{\gamma}{B_n^*} - \frac{1}{4S_{n+1}^*}, \\ n = 1, 2, 3, \ldots. & \end{array}\right\} \qquad (3.37)$$

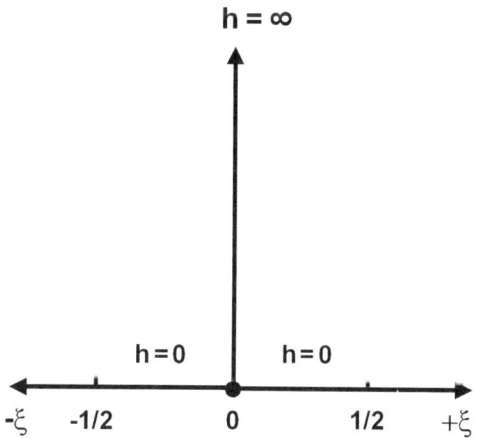

Fig. 3.7. Delta-like law of pulsations of the THTC

Substituting these relations into (3.26) and finding its limiting form at $\gamma \to 0$, one can obtain an uncertainty of the following kind:

$$\varepsilon = \frac{\gamma}{1 - (S_1^{-1} + S_1^{*-1})/4} \sim \frac{0}{0}. \tag{3.38}$$

Solving this uncertainty by means of the L'Hospital's rule [21] gives

$$\varepsilon = \left(2\langle \bar{h} \rangle \sum_{n=1}^{\infty} \frac{F_n}{F_n^2 + \Phi_n^2}\right)^{-1}. \tag{3.39}$$

One can render to the system of (3.26) at $b \to 1$ a more evident form, having rewritten it using other mathematical designations:

$$\left. \begin{array}{c} \dfrac{h_m}{2h_{\min}} = \dfrac{1}{1 - (S_1^{-1} + S_1^{*-1})/4}, \\[2mm] S_1 = 1 + \dfrac{2h_{\min}}{B_1} - \dfrac{1}{4S_2}, \quad S_1^* = 1 + \dfrac{2h_{\min}}{B_1^*} - \dfrac{1}{4S_2^*}, \\[2mm] S_2 = 1 + \dfrac{2h_{\min}}{B_2} - \dfrac{1}{4S_3}, \quad S_2^* = 1 + \dfrac{2h_{\min}}{B_2^*} - \dfrac{1}{4S_3^*}, \\[2mm] \vdots \\[2mm] S_n = 1 + \dfrac{2h_{\min}}{B_n} - \dfrac{1}{4S_{n+1}}, \quad S_n^* = 1 + \dfrac{2h_{\min}}{B_n^*} - \dfrac{1}{4S_{n+1}^*}, \\[2mm] n = 1, 2, 3, \ldots. \end{array} \right\} \tag{3.40}$$

An analysis of the resulting relations at $b \to 1$ shows that the EHTC ceases to depend on the ATHTC and is determined by a minimal value of the THTC $h_{\min}$ over a period. This effect could not be foreseen beforehand. From the physical point of view, it means that at pulsing variation of the heat transfer intensity the average level of heat transfer is adjusted to the minimal values over the period. In real physical applications, the delta-like law of oscillations of the THTC investigated here can appear at harmonic oscillations of a liquid film thickness on a heated surface described by the function $\delta_f = \langle \delta_f \rangle [1 + \cos(\xi)]$. Then at certain moments of time one can inevitably have: $\delta_f = 0, h = k_f / \delta_f = \infty$. We have investigated in this section the case of the delta-like oscillations acquired by the heat transfer intensity due to the limiting transition $b \to 1$ in the general solution for the inverse harmonic law of oscillations of the THTC. It is also interesting to study this limiting case, having initially specified the THTC as a periodic delta-like function.

3.4 Delta-Like Law of Oscillations

The delta-function $\delta(y - y_0)$ is defined by the following relations [21]

$$\left.\begin{array}{l} y = y_0 : \delta(y - y_0) = \infty, \\ y \neq y_0 : \delta(y - y_0) = 0, \\ \int\limits_{-\infty}^{\infty} \delta(y - y_0) \, dy = 1, \\ \int\limits_{-\infty}^{\infty} f(y) \delta(y - y_0) \, dy = f(y_0). \end{array}\right\} \quad (3.41)$$

Let us define a delta – function $\delta(\xi_0)$ over the period $0 \leq \xi \leq 2\pi$ by the relations similar to the system of (3.41)

$$\left.\begin{array}{l} \xi = \xi_0 : \delta(\xi - \xi_0) = \infty, \\ \xi \neq \xi_0 : \delta(\xi - \xi_0) = 0, \\ \int\limits_{0}^{2\pi} \delta(\xi - \xi_0) \, d\xi = 1, \\ \int\limits_{0}^{2\pi} f(\xi) \delta(\xi - \xi_0) \, d\xi = f(\xi_0) \end{array}\right\}. \quad (3.42)$$

Let us preset the THTC as a periodic delta-function

$$h(\xi) = \langle h \rangle [1 + \psi(\xi)] = 2\pi \langle h \rangle \delta(\xi - \xi_0). \quad (3.43)$$

Let us further write the expressions for the periodic components of the temperature and heat flux functions on a heat transfer surface as

$$\theta = \sum_{n=1}^{\infty} [R_n \cos(n\xi) - I_n \sin(n\xi)], \quad (3.44)$$

$$\theta^{\bullet} = \sum_{n=1}^{\infty} [(F_n R_n - \Phi_n I_n) \cos(n\xi) - (F_n I_n + \Phi_n R_n) \sin(n\xi)]. \quad (3.45)$$

Expressions (3.44) and (3.45) become equivalent to (3.1) and (3.2). Let us substitute values $\theta, \theta^{\bullet}, \psi$ in the BC (2.29)

$$2\pi\delta(\xi - \xi_0) \left[1 + \sum_{n=1}^{\infty} [R_n \cos(n\xi) - I_n \sin(n\xi)]\right]$$

$$= \varepsilon + \frac{1}{\langle h \rangle} \sum_{n=1}^{\infty} [(\Phi_n I_n - F_n R_n) \cos(n\xi) + (F_n I_n + \Phi_n R_n) \sin(n\xi)]. \quad (3.46)$$

In order to solve (3.46), let us apply Galerkin's method [22]. First, one needs to average both parts of (3.46) via integration over the period with the weights $1, \sin(n\xi), \cos(n\xi)$, respectively

$$\left.\begin{array}{l}\varepsilon = 1 + \sum_{n=1}^{\infty} [R_n \cos(n\xi_0) - I_n \sin(n\xi_0)], \\ 2\left\langle \bar{h} \right\rangle \varepsilon \sin(n\xi_0) = F_n I_n + \Phi_n R_n, \\ 2\left\langle \bar{h} \right\rangle \varepsilon \cos(n\xi_0) = \Phi_n I_n - F_n R_n.\end{array}\right\} \quad (3.47)$$

The solution of the system of (3.47) looks like

$$\varepsilon = \left(1 + 2\left\langle \bar{h} \right\rangle \sum_{n=1}^{\infty} \frac{F_n}{F_n^2 + \Phi_n^2}\right)^{-1}. \quad (3.48)$$

Galerkin's method that belongs to the so-called direct methods in the calculus of variations is traditionally considered an approximate technique [17]. The reason for such a classification is the approximation of an exact solution with a finite set of basic functions performed within the framework of this method. However, in our case, we preset the field of the temperature oscillations in a form of infinite Fourier series (3.44). This effectively means that from the very beginning we search for an exact analytical solution of the problem. From the point of view of the ideology of the Galerkin's method [22], this means a use of the full system of basic functions. For this reason, (3.48) represents an exact analytical solution. As shown in Appendix D, the series written in right-hand sides of (3.39) and (3.48) are always diverging. Thus, the used procedure provides a generalization of the asymptotical form of the extremely asymmetric law of oscillations $b \to 1: \varepsilon \to 0$. It is interesting to note that for the case of a semi-infinite body $(\bar{\delta} \to \infty)$ dependence (3.48) transfers into dependence for $\varepsilon\left(\left\langle \bar{h} \right\rangle\right)$, which is described by Kroeneker symbol

$$\left.\begin{array}{l}\left\langle \tilde{h} \right\rangle = 0 : \varepsilon = 1 \\ \left\langle \tilde{h} \right\rangle \neq 0 : \varepsilon = 0\end{array}\right\}. \quad (3.49)$$

Generally saying, the analysis of the inverse harmonic type of oscillations of the heat transfer intensity performed in this section repeats (in the sense of its mathematical content and used practical tools) the case of the harmonic oscillations considered above. The basic difference of the present analysis in comparison with that given in the previous section consists in the fact that the growth of the amplitude b in the present case results in an amplification of the level of oscillations' asymmetry in a process (Fig. 3.7). As shown above, at $b = 1$ we obtain extremely asymmetric (delta-like) law of oscillations of the THTC. Its peculiarity consists in the abnormal behavior of the conjugated system: $h_m = 0, \varepsilon = 0$ correspond to finite values of $\langle h \rangle$. This effectively means that at a fixed average heat flux supplied to an external surface of a body, the temperature difference "body – fluid" on an internal side of a body at $b \to 1$ grows infinitely. A common feature uniting harmonic and inverse harmonic types of oscillations of the THTC is their smooth character. According to a classification of periodic oscillations in the sense of their complexity, the next more complicated form is that described by a step function.

3.5 Step Law of Oscillations

The general solution for symmetric step oscillations. As shown above, a step behavior of the THTC variation is inherent to a slug regime of a two-phase fluid flow in a pipe (intermittent flow of steam and liquid volumes), and also to a wave flow of a liquid film for significant mass flow-rates of a liquid (movement of a liquid in a form of drops rolling down over a wet surface). In the present section, a special case of the step oscillations in a form of a symmetric step function (Fig. 3.8) is considered. The periodic part of the THTC is preset in this case by means of the following relations

$$\left. \begin{array}{l} 0 \leq \xi \leq \pi : \psi = b \\ \pi \leq \xi \leq 2\pi : \psi = -b \end{array} \right\} \quad (3.50)$$

and is expressed as a Fourier series

$$\psi = \frac{4b}{\pi} \sum_{n=1}^{\infty} \frac{\sin\left[(2n-1)\xi\right]}{2n-1}, \quad (3.51)$$

or, in a complex form of notation, as the following series

$$\psi = -\frac{2bi}{\pi} \sum_{n=1}^{\infty} \frac{\exp\left[(2n-1)i\xi\right] - \exp\left[-(2n-1)i\xi\right]}{2n-1}. \quad (3.52)$$

The solution procedure involves a substitution of expressions for the oscillations of temperatures, heat fluxes and the THTC into the BC (2.29) and a consecutive comparison of coefficients at identical exponential terms $\exp\left[\pm(2n-1)i\xi\right]$. This entails the following correlation for the FC with odd eigenvalues A_n, A_n^*

$$\varepsilon_1 = 1 + \frac{4bi}{\pi} \sum_{n=1}^{\infty} \frac{A_{2n-1} - A_{2n-1}^*}{2n-1}. \quad (3.53)$$

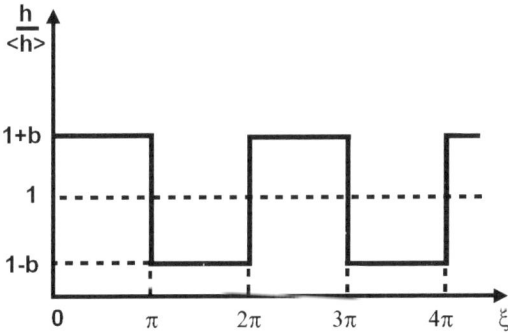

Fig. 3.8. Step law of pulsations of the THTC

56   3 Solution of Characteristic Problems

Resulting expressions for the eigenvalues A_n, A_n^* include double infinite series and are not presented here because of their cumbersome form. For the considered case of a step function $\psi(\xi)$, it is unfortunately impossible to derive an analytical solution similar to the solutions obtained above for the cases of harmonic and inverse harmonic functions $\psi(\xi)$. It is possible to show that the final system of algebraic equations is equivalent to a system of the Fredholm integral equations of the second kind. One can also obtain its approximate solution using, for example, an iterative method of Neumann [8]. An analysis of the first approximation shows that the expressions containing double series do not contribute to the first two pairs of the odd eigenvalues $A_1, A_1^*; A_3, A_3^*$. We will use this property of the first approximation below, while deriving an analytical solution. If one uses a notation of the BC in the form of (2.30), this results in an alternative expression for the FC

$$\varepsilon_2 = 1 - b^2 - \frac{2bi}{\pi \langle \bar{h} \rangle} \sum_{n=1}^{\infty} \frac{B_{2n-1}^* A_{2n-1} - B_{2n-1}^* A_{2n-1}^*}{2n - 1}. \tag{3.54}$$

The aforementioned property of the eigenvalues (the one following from the iterative method of Neumann) remains in force in (3.54) as well.

Analytical solution. Let us apply the following algorithm for an analytical solution of the problem. The property of the first pairs of eigenvalues $A_1, A_1^*; A_3, A_3^*$ proved by means of the iterative method of Neumann is valid for any number n. From this property, a recurrent formula follows immediately: $A_{2n} = 0, A_{2n}^* = 0, n = 1, 2, 3, \ldots$. Further, both alternative expressions for the FC [(3.53) and (3.54)] are considered separately. Expressions for the eigenvalues following from (3.53) look like

$$A_{2n-1} = \frac{2bi}{\pi} \frac{1}{1 + B_{2n-1}/\langle \bar{h} \rangle}, \quad A_{2n-1}^* = \frac{2bi}{\pi} \frac{1}{1 + B_{2n-1}^*/\langle \bar{h} \rangle}. \tag{3.55}$$

Substituting them into (3.53), one can obtain the first solution for the FC

$$\varepsilon_1 = 1 - \frac{8b^2 \langle \bar{h} \rangle}{\pi^2} \sum_{n=1}^{\infty} \frac{1}{(2n-1)^2} \frac{\langle \bar{h} \rangle + F_{2n-1}}{\left(\langle \bar{h} \rangle + F_{2n-1} \right)^2 + \Phi_{2n-1}^2}. \tag{3.56}$$

The following expressions for the eigenvalues can be further derived from (3.7)–(3.9), (3.51):

$$A_{2n-1} = \frac{2b\varepsilon i}{\pi(2n-1)} \frac{1}{1 - b^2 + B_{2n-1}/\langle \bar{h} \rangle}, \tag{3.57}$$

$$A_{2n-1}^* = \frac{2b\varepsilon i}{\pi(2n-1)} \frac{1}{1 - b^2 + B_{2n-1}^*/\langle \bar{h} \rangle}. \tag{3.58}$$

Substituting them into (3.54), one can obtain the second solution for the FC

$$\varepsilon_2 = (1-b^2) \left\{ 1 - \frac{8b^2}{\pi^2} \sum_{n=1}^{\infty} \frac{1}{(2n-1)^2} \frac{F_{2n-1}\left[(1-b^2)\langle \bar{h} \rangle + F_{2n-1}\right] + \Phi_{2n-1}^2}{(1-b^2)\left(\langle \bar{h} \rangle + F_{2n-1}\right)^2 + \Phi_{2n-1}^2} \right\}^{-1}. \tag{3.59}$$

3.5 Step Law of Oscillations

An analysis of expressions (3.56) and (3.59) results in the following conclusions.

- Each of the alternative solutions $\varepsilon_1, \varepsilon_2$ has four analytical asymptotical forms: $\langle \bar{h} \rangle \to 0$; $\langle \bar{h} \rangle \to \infty$; $\delta \to 0$ for the TBC $\vartheta_0 = $ const; $\delta \to 0$ for the TBC $q_0 = $ const.
- These asymptotical forms can be presented conventionally as two groups: a limiting case of a weak conjugation ("a" and "c"); a limiting case of a strong conjugation ("b" and "d").
- All four asymptotical forms coincide with the exact solutions (obtained as the first approximation using the Neumann's method).
- All four asymptotical forms of the alternative solutions $\varepsilon_1 - \varepsilon_2$ coincide among themselves.

On the basis of the performed asymptotic analysis, it is possible to assume with a high degree of confidence that the difference $\varepsilon_1 - \varepsilon_2$ gives a maximal error at calculation using one of the alternative relations, namely, (3.56) or (3.59). Results of a calculation of the FC for the step law of oscillations of the THTC are shown in Tables 3.5–3.10 and in Figs. 3.9–3.14. Asymptotical forms of the obtained approximate solutions (to within two terms of an expansion in a small parameter) are given below.

Asymptotical solution at $b \to 0$. This asymptotical form is natural. At disappearance of the heat transfer intensity oscillations, a distinction between the ATHTC and EHTC also vanishes, so that in this case one has $\varepsilon = 1$.

Asymptotical solution at $\langle \bar{h} \rangle \to 0$. This asymptotical form can be written down as the first two terms of a Taylor series in a small parameter $\langle \bar{h} \rangle$:

$$\varepsilon = 1 - \frac{8b^2 \langle \bar{h} \rangle}{\pi^2} \sum_{n=1}^{\infty} \frac{1}{(2n-1)^2} \frac{F_{2n-1}}{F_{2n-1}^2 + \Phi_{2n-1}^2}. \tag{3.60}$$

In the case of semi-infinite bodies $\delta \to \infty$, an influence of the wall thickness degenerates. Therefore, one can write for a time-dependent problem

$$\varepsilon = 1 - \frac{4\sqrt{2}b^2 \langle \tilde{h} \rangle}{\pi^2} \sum_{n=1}^{\infty} \frac{1}{(2n-1)^{5/2}}, \quad \langle \tilde{h} \rangle = \frac{\langle h \rangle \sqrt{\alpha \tau_0}}{k} \tag{3.61}$$

and

$$\varepsilon = 1 - \frac{8b^2 \langle \bar{h} \rangle}{\pi^2} \sum_{n=1}^{\infty} \frac{1}{(2n-1)^3}, \quad \langle \bar{h} \rangle = \frac{\langle h \rangle Z_0}{k} \tag{3.62}$$

for a spatial problem, respectively.

Asymptotical solution at $\langle \bar{h} \rangle \to \infty$. This asymptotical form can be also presented as the first two members of the Taylor series in a small parameter $\langle \bar{h} \rangle^{-1}$:

$$\varepsilon = 1 - b^2 + \frac{8b^2}{\pi^2 \langle \bar{h} \rangle} \sum_{n=1}^{\infty} \frac{F_{2n-1}}{(2n-1)^2}. \tag{3.63}$$

3 Solution of Characteristic Problems

Table 3.5. Step pulsations

$\langle \bar{h} \rangle$	B					
	1.0	0.95	0.90	0.80	$2/\pi$	0.50
0.1	0.9222	0.9298	0.9370	0.9502	0.9685	0.9806
0.2	0.8569	0.8708	0.8841	0.9084	0.9420	0.9642
0.3	0.8012	0.8206	0.8390	0.8728	0.9194	0.9503
0.4	0.7532	0.7772	0.8001	0.8420	0.9000	0.9383
0.5	0.7112	0.7394	0.7661	0.8152	0.8830	0.9278
0.6	0.6743	0.7060	0.7362	0.7915	0.8680	0.9186
0.7	0.6414	0.6764	0.7096	0.7705	0.8547	0.9104
0.8	0.6121	0.6499	0.6858	0.7517	0.8428	0.9030
0.9	0.5856	0.5260	0.6643	0.7348	0.8320	0.8964
1.0	0.5616	0.6043	0.6449	0.7194	0.8223	0.8904
2.0	0.4050	0.4630	0.5180	0.6192	0.7588	0.8512
3.0	0.3220	0.3881	0.4508	0.5661	0.7252	0.8305
4.0	0.2698	0.3410	0.4085	0.5327	0.7041	0.8174
5.0	0.2335	0.3083	0.3792	0.5095	0.6894	0.8084
6.0	0.2067	0.2840	0.3574	0.4923	0.6785	0.8017
7.0	0.1860	0.2653	0.3406	0.4790	0.6701	0.7965
8.0	0.1694	0.2504	0.3272	0.4684	0.6634	0.7923
9.0	0.1558	0.2381	0.3162	0.4597	0.6578	0.7890
10	0.1444	0.2278	0.3070	0.4524	0.6532	0.7861
20	0.08597	0.1751	0.2596	0.4150	0.6296	0.7715
30	0.06262	0.1540	0.2407	0.4001	0.6201	0.7657
40	0.04976	0.1424	0.2303	0.3918	0.6149	0.7624
50	0.04153	0.1350	0.2236	0.3866	0.6116	0.7604
60	0.03577	0.1298	0.2190	0.3829	0.6092	0.7589
70	0.03150	0.1259	0.2155	0.3802	0.6075	0.7579
80	0.02819	0.1229	0.2128	0.3780	0.6061	0.7570
90	0.02554	0.1206	0.2107	0.3764	0.6051	0.7564
100	0.02338	0.1186	0.2089	0.3750	0.6042	0.7559
200	0.01296	0.1092	0.2005	0.3683	0.6000	0.7532
∞	0	0.09750	0.1900	0.3600	0.5947	0.7500

Spatial problem for a semi-infinite body. Values of the factor of conjugation

In the case semi-infinite bodies $\delta \to \infty$, one can derive for a time-dependent problem

$$\varepsilon = 1 - b^2 + \frac{4\sqrt{2}b^2}{\pi^2 \langle \tilde{h} \rangle} \sum_{n=1}^{\infty} \frac{1}{(2n-1)^{3/2}} \qquad (3.64)$$

and

$$\varepsilon = 1 - b^2 + \frac{8b^2}{\pi^2 \langle \bar{h} \rangle} \sum_{n=1}^{\infty} \frac{1}{2n-1} \qquad (3.65)$$

for a spatial problem.

Calculation of the sums of infinite series. For a calculation of the sums of the infinite series in (3.61)–(3.65), one should rearrange these equations as

$$s(N) = \sum_{n=1}^{\infty} \frac{1}{(2n-1)^N} = 1 - \frac{1}{2^N} \varsigma(N), \qquad (3.66)$$

3.5 Step Law of Oscillations

Table 3.6. Step pulsations

$\langle \tilde{h} \rangle$	B					
	1.0	0.95	0.90	0.80	$2/\pi$	0.50
0.1	0.9372	0.9433	0.9491	0.9598	0.9745	0.9843
0.2	0.8769	0.8889	0.9003	0.9212	0.9501	0.9692
0.3	0.8206	0.8381	0.8547	0.8852	0.9273	0.9551
0.4	0.7687	0.7912	0.8126	0.8520	0.9062	0.9422
0.5	0.7213	0.7485	0.7743	0.8216	0.8870	0.9303
0.6	0.6782	0.7096	0.7394	0.7941	0.8696	0.9196
0.7	0.6392	0.6744	0.7077	0.7691	0.8538	0.9098
0.8	0.6038	0.6424	0.6790	0.7464	0.8394	0.9009
0.9	0.5716	0.6133	0.6530	0.7258	0.8264	0.8929
1.0	0.5423	0.5869	0.6292	0.7070	0.8145	0.8856
2.0	0.3538	0.4168	0.4765	0.5864	0.7381	0.8384
3.0	0.2604	0.3325	0.4010	0.5267	0.7003	0.8151
4.0	0.2056	0.2831	0.3565	0.4916	0.6780	0.8014
5.0	0.1697	0.2506	0.3275	0.4686	0.6635	0.7924
6.0	0.1444	0.2278	0.3070	0.4524	0.6532	0.7861
7.0	0.1256	0.2109	0.2918	0.4404	0.6456	0.7814
8.0	0.1112	0.1978	0.2801	0.4312	0.6398	0.7778
9.0	0.09971	0.1875	0.2708	0.4238	0.6351	0.7749
10	0.09037	0.1791	0.2632	0.4178	0.6313	0.7726
20	0.04668	0.1396	0.2278	0.3899	0.6136	0.7617
30	0.03152	0.1260	0.2155	0.3802	0.6075	0.7579
40	0.02384	0.1190	0.2093	0.3753	0.6044	0.7560
50	0.01920	0.1148	0.2056	0.3723	0.6025	0.7548
60	0.01610	0.1120	0.2030	0.3703	0.6012	0.7540
70	0.01388	0.1100	0.2012	0.3689	0.6003	0.7535
80	0.01221	0.1085	0.1999	0.3678	0.5997	0.7530
90	0.01091	0.1074	0.1988	0.3670	0.5991	0.7527
100	0.009874	0.1064	0.1980	0.3663	0.5987	0.7525
200	0.005188	0.1022	0.1942	0.3633	0.5968	0.7513
∞	0	0.09750	0.1900	0.3600	0.5947	0.7500

Time-dependent problem for a semi-infinite body. Values of the factor of conjugation

where $\varsigma(N) = \sum_{n=1}^{\infty} \frac{1}{n^N}$ is the Riemann zeta-function, and $\varsigma(1) = \infty$ [23]. In view of the tabulated values of the Riemann zeta-function, expansions (3.61), (3.62), and (3.64) can be also rewritten

$$\varepsilon = 1 - 0.6330 b^2 \langle \tilde{h} \rangle, \qquad (3.61a)$$

$$\varepsilon = 1 - 0.8526 b^2 \langle \bar{h} \rangle, \qquad (3.62a)$$

$$\varepsilon = 1 - b^2 + \frac{0.9679}{\langle \tilde{h} \rangle}. \qquad (3.64a)$$

Since the series in the right-hand side of (3.65) is diverging, the asymptotical solution under consideration is not analytical. To investigate its properties, we intend to use the Euler–MacLaurin summation formula [24]

Table 3.7. Step pulsations

$\tilde{\delta}$	$\langle \bar{h} \rangle$									
	0.1	0.2	0.3	0.5	0.7	1.0	2.0	3.0	5.0	7.0	10
0.01	0.9990	0.9980	0.9970	0.9950	0.9931	0.9901	0.9805	0.9710	0.9526	0.9348	0.9094
0.02	0.9980	0.9960	0.9940	0.9901	0.9862	0.9804	0.9616	0.9434	0.9092	0.8773	0.8335
0.03	0.9970	0.9940	0.9911	0.9852	0.9794	0.9709	0.9434	0.9175	0.8696	0.8265	0.7693
0.04	0.9960	0.9921	0.9881	0.9804	0.9728	0.9615	0.9259	0.8928	0.8333	0.7812	0.7143
0.05	0.9950	0.9901	0.9852	0.9756	0.9662	0.9524	9.9091	0.8696	0.8000	0.7408	0.6667
0.06	0.9940	0.9881	0.9823	0.9709	0.9597	0.9436	0.8929	0.8475	0.7693	0.7043	0.6250
0.07	0.9930	0.9862	0.9794	0.9662	0.9533	0.9346	0.8772	0.8265	0.7408	0.6712	0.5883
0.08	0.9921	0.9843	0.9766	0.9616	0.9470	0.9260	0.8621	0.8066	0.7144	0.6412	0.5557
0.09	0.9911	0.9823	0.9737	0.9570	0.9408	0.9175	0.8476	0.7876	0.6898	0.6137	0.5265
0.10	0.9901	0.9804	0.9709	0.9524	0.9347	0.9092	0.8335	0.7695	0.6670	0.5886	0.5003
0.20	0.9805	0.9618	0.9438	0.9097	0.8780	0.8344	0.7158	0.6268	0.5019	0.4185	0.3350
0.30	0.9714	0.9444	0.9188	0.8716	0.8290	0.7724	0.6292	0.5308	0.4043	0.3265	0.2534
0.40	0.9628	0.9282	0.8960	0.8379	0.7869	0.7210	0.5637	0.4627	0.3406	0.2695	0.2052
0.50	0.9547	0.9134	0.8755	0.8083	0.7507	0.6782	0.5130	0.4125	0.2963	0.2312	0.1739
0.60	0.9474	0.9001	0.8572	0.7825	0.7199	0.6427	0.4732	0.3744	0.2641	0.2040	0.1520
0.70	0.9409	0.8883	0.8412	0.7605	0.6939	0.6132	0.4417	0.3450	0.2400	0.1840	0.1362
0.80	0.9352	0.8781	0.8275	0.7418	0.6721	0.5888	0.4165	0.3221	0.2215	0.1688	0.1244
0.90	0.9304	0.8696	0.8160	0.7262	0.6540	0.5689	0.3964	0.3040	0.2072	0.1572	0.1153
1.0	0.9266	0.8627	0.8068	0.7136	0.6394	0.5528	0.3803	0.2896	0.1960	0.1481	0.1083
2.0	0.9290	0.8623	0.8013	0.6967	0.6128	0.5163	0.3344	0.2462	0.1608	0.1193	0.08598
3.0	0.9376	0.8772	0.8205	0.7202	0.6372	0.5393	0.3499	0.2569	0.1670	0.1235	0.08878
4.0	0.9377	0.8778	0.8217	0.7227	0.6406	0.5437	0.3548	0.2613	0.1703	0.1261	0.09072
5.0	0.9372	0.8770	0.8207	0.7215	0.6395	0.5426	0.3542	0.2609	0.1702	0.1260	0.09070
∞	0.9372	0.8769	0.8206	0.7213	0.6392	0.5423	0.3538	0.2609	0.1697	0.1256	0.09037

Time-dependent problem. TBC $\vartheta_0 = $ const. Values of the factor of conjugation

$$\sum_{n=0}^{N} f(n) = \int_{0}^{N} f(y)\,dy + \frac{f(0)+f(N)}{2} + \sum_{n=1}^{\infty} \frac{D_{2n}}{(2n)!}\left[f^{(2n-1)}(N) - f^{(2n-1)}(0)\right]$$
$$+ \frac{ND_{2N+2}}{(2N+2)!} f^{(2N+2)}(\kappa N). \tag{3.67}$$

Here D_n are Bernoulli numbers [24], $n = 1, 2, 3, \ldots, N = 1, 2, 3, \ldots, 0 \leq \kappa \leq 1$.

Let us simplify solution (3.56) for a particular case of semi-infinite bodies $\delta \to \infty$

$$\varepsilon = 1 - \frac{8b^2 \langle \bar{h} \rangle}{\pi^2} \sum_{n=1}^{\infty} \frac{1}{(2n-1)^2 (2n-1+\langle \bar{h} \rangle)}, \tag{3.68}$$

that, in view of the sum of a tabulated series [24]

$$\sum_{n=1}^{\infty} \frac{1}{(2n-1)^2} = \frac{8}{\pi^2}, \tag{3.69}$$

3.5 Step Law of Oscillations   61

Table 3.8. Step pulsations

$\langle \tilde{h} \rangle$

$\tilde{\delta}$	0.1	0.2	0.3	0.5	0.7	1.0	2.0	3.0	5.0	7.0	10
0.01	0.06361	0.03177	0.02116	0.01267	0.009036	0.006306	0.003124	0.002063	0.001215	0.0008515	0.000579
0.02	0.1273	0.06361	0.04238	0.02541	0.01813	0.01267	0.006308	0.004186	0.002489	0.001762	0.001217
0.03	0.1909	0.09544	0.06361	0.03814	0.02723	0.01904	0.009491	0.006308	0.003762	0.002671	0.001854
0.04	0.2544	0.1273	0.08483	0.05087	0.03632	0.02541	0.01268	0.008431	0.005035	0.003581	0.002491
0.05	0.3171	0.1591	0.1060	0.06361	0.04542	0.03177	0.01586	0.01055	0.006309	0.004491	0.003128
0.06	0.3778	0.1909	0.1273	0.07634	0.05451	0.03814	0.01904	0.01268	0.007582	0.005400	0.003764
0.07	0.4356	0.2227	0.1485	0.08907	0.06361	0.04451	0.02222	0.01480	0.008856	0.006310	0.004401
0.08	0.4894	0.2544	0.1697	0.1018	0.07270	0.05087	0.02541	0.01692	0.01013	0.007219	0.005038
0.09	0.5387	0.2859	0.1909	0.1145	0.08180	0.05724	0.02859	0.01904	0.01140	0.008129	0.005674
0.10	0.5833	0.3170	0.2121	0.1273	0.09089	0.06361	0.03178	0.02116	0.01268	0.009039	0.006311
0.20	0.8312	0.5817	0.4161	0.2544	0.1818	0.1273	0.06361	0.04239	0.02541	0.01813	0.01268
0.30	0.9100	0.7377	0.5790	0.3768	0.2722	0.1909	0.09544	0.06361	0.03814	0.02723	0.01904
0.40	0.9402	0.8199	0.6891	0.4842	0.3590	0.2541	0.1273	0.08483	0.05088	0.03633	0.02541
0.50	0.9535	0.8638	0.7587	0.5700	0.4372	0.3153	0.1591	0.1060	0.06361	0.04542	0.03178
0.60	0.9596	0.8878	0.8017	0.6342	0.5032	0.3721	0.1907	0.1273	0.07634	0.05452	0.03815
0.70	0.9621	0.9010	0.8281	0.6802	0.5802	0.4223	0.2217	0.1484	0.08907	0.06361	0.04451
0.80	0.9627	0.9081	0.8440	0.7120	0.7334	0.4646	0.2513	0.1691	0.1017	0.07269	0.05088
0.90	0.9623	0.9114	0.8532	0.7334	0.7334	0.4985	0.2785	0.1890	0.1142	0.08169	0.05721
1.0	0.9613	0.9124	0.8581	0.7473	0.7473	0.5247	0.3025	0.2076	0.1262	0.09048	0.06343
2.0	0.9446	0.8903	0.8386	0.7452	0.7452	0.5692	0.3750	0.2764	0.1800	0.1330	0.09554
3.0	0.9368	0.8767	0.8207	0.7225	0.7225	0.5454	0.3579	0.2644	0.1729	0.1282	0.09238
4.0	0.9367	0.8761	0.8195	0.7200	0.7200	0.5410	0.3530	0.2600	0.1696	0.1257	0.09044
5.0	0.9372	0.8769	0.8205	0.7212	0.7212	0.5421	0.3540	0.2604	0.1698	0.1257	0.09047
∞	0.9372	0.8769	0.8206	0.7213	0.6392	0.5423	0.3538	0.2607	0.1697	0.1256	0.09037

Time-dependent problem. TBC q_0 = const. Values of the factor of conjugation

Table 3.9. Step pulsations

$\bar{\delta}$	$\langle \bar{h} \rangle$									
	0.1	0.2	0.3	0.5	0.7	1.0	2.0	3.0	5.0	7.0	10
0.01	0.9990	0.9980	0.9970	0.9950	0.9931	0.9901	0.9805	0.9710	0.9526	0.9348	0.9094
0.02	0.9980	0.9960	0.9941	0.9902	0.9863	0.9805	0.9618	0.9438	0.9097	0.8770	0.8344
0.03	0.9970	0.9941	0.9912	0.9854	0.9796	0.9712	0.9440	0.9182	0.8708	0.8281	0.7713
0.04	0.9961	0.9922	0.9883	0.9807	0.9731	0.9621	0.9269	0.8942	0.8354	0.7839	0.7176
0.05	0.9951	0.9903	0.9855	0.9760	0.9668	0.9532	0.9106	0.8716	0.8030	0.7445	0.6712
0.06	0.9942	0.9884	0.9827	0.9715	0.9605	0.9445	0.8949	0.8503	0.7734	0.7093	0.6310
0.07	0.9932	0.9865	0.9799	0.9670	0.9544	0.9361	0.8800	0.8302	0.7460	0.6775	0.5953
0.08	0.9923	0.9847	0.9772	0.9626	0.9484	0.9279	0.8656	0.8112	0.7208	0.6488	0.5644
0.09	0.9914	0.9829	0.9745	0.9582	0.9425	0.9199	0.8518	0.7932	0.6975	0.6226	0.5366
0.10	0.9904	0.9811	0.9719	0.9540	0.9368	0.9121	0.8385	0.7761	0.6758	0.5988	0.5117
0.20	0.9817	0.9641	0.9472	0.9150	0.8851	0.8437	0.7304	0.6449	0.5234	0.4412	0.3580
0.30	0.9738	0.9491	0.9256	0.8821	0.8426	0.7899	0.6549	0.5606	0.4370	0.3594	0.2850
0.40	0.9667	0.9357	0.9068	0.8542	0.8076	0.7470	0.5998	0.5029	0.3826	0.3104	0.2434
0.50	0.9604	0.9240	0.8904	0.8305	0.7787	0.7126	0.5585	0.4617	0.3459	0.2785	0.2171
0.60	0.9547	0.9137	0.8763	0.8106	0.7547	0.6848	0.5271	0.4314	0.3200	0.2564	0.1993
0.70	0.9498	0.9048	0.8642	0.7938	0.7348	0.6623	0.5027	0.4085	0.3010	0.2406	0.1867
0.80	0.9455	0.8971	0.8538	0.7797	0.7184	0.6439	0.4836	0.3910	0.2868	0.2289	0.1775
0.90	0.9418	0.8905	0.8451	0.7679	0.7048	0.6290	0.4685	0.3773	0.2759	0.2200	0.1706
1.0	0.9386	0.8849	0.8377	0.7581	0.6936	0.6168	0.4564	0.3666	0.2675	0.2132	0.1653
2.0	0.9246	0.8610	0.8065	0.7179	0.6488	0.5692	0.4119	0.3280	0.2381	0.1896	0.1473
3.0	0.9226	0.8575	0.8020	0.7123	0.6426	0.5628	0.4061	0.3231	0.2345	0.1868	0.1452
4.0	0.9223	0.8570	0.8014	0.7115	0.6418	0.5620	0.4054	0.3224	0.2340	0.1864	0.1449
5.0	0.9222	0.8570	0.8013	0.7114	0.6416	0.5618	0.4053	0.3224	0.2339	0.1864	0.1448
∞	0.9222	0.8569	0.8012	0.7112	0.6414	0.5616	0.4050	0.3220	0.2335	0.1860	0.1444

Spatial problem. TBC $\vartheta_0 = $ const. Values of the factor of conjugation

3.5 Step Law of Oscillations   63

Table 3.10. Step pulsations

$\bar{\delta}$	$\langle \tilde{h} \rangle$										
	0.1	0.2	0.3	0.5	0.7	1.0	2.0	3.0	5.0	7.0	10
0.01	0.2012	0.1422	0.1161	0.08990	0.07595	0.06350	0.04482	0.03653	0.02820	0.2376	0.01970
0.02	0.2841	0.2011	0.1642	0.1271	0.1073	0.08970	0.06322	0.05146	0.03963	0.03332	0.02766
0.03	0.3462	0.2461	0.2010	0.1555	0.1313	0.1097	0.07722	0.06278	0.04824	0.04047	0.03350
0.04	0.3968	0.2838	0.2319	0.1794	0.1515	0.1265	0.08894	0.07222	0.05538	0.04636	0.03828
0.05	0.4393	0.3166	0.2590	0.2005	0.1692	0.1412	0.09918	0.08045	0.06156	0.05145	0.04238
0.06	0.4758	0.3457	0.2834	0.2194	0.1852	0.1545	0.1084	0.08781	0.06706	0.05596	0.04600
0.07	0.5076	0.3720	0.3055	0.2368	0.1998	0.1667	0.1168	0.09451	0.07205	0.06002	0.04924
0.08	0.5356	0.3960	0.3260	0.2529	0.2133	0.1779	0.1245	0.1007	0.07673	0.06374	0.05219
0.09	0.5604	0.4179	0.3449	0.2679	0.2260	0.1885	0.1317	0.1064	0.08087	0.06717	0.05490
0.10	0.5827	0.4381	0.3626	0.2820	0.2380	0.1984	0.1385	0.1118	0.08482	0.07037	0.05742
0.20	0.7207	0.5793	0.4929	0.3909	0.3316	0.2765	0.1916	0.1535	0.1149	0.09431	0.07599
0.30	0.7876	0.6599	0.5740	0.4649	0.3975	0.3327	0.2299	0.1831	0.1358	0.1106	0.08837
0.40	0.8264	0.7114	0.6292	0.5187	0.4472	0.3763	0.2602	0.2064	0.1520	0.1231	0.09774
0.50	0.8514	0.7466	0.6685	0.5591	0.4857	0.4109	0.2849	0.2255	0.1652	0.1333	0.1053
0.60	0.8685	0.7718	0.6974	0.5901	0.5159	0.4388	0.3054	0.2415	0.1762	0.1417	0.1116
0.70	0.8808	0.7903	0.7192	0.6142	0.5400	0.4614	0.3225	0.2549	0.1856	0.1489	0.1168
0.80	0.8898	0.8043	0.7360	0.6332	0.5592	0.4797	0.3367	0.2662	0.1935	0.1550	0.1213
0.90	0.8966	0.8151	0.7490	0.6482	0.5746	0.4947	0.3586	0.2757	0.2002	0.1601	0.1252
1.0	0.9018	0.8234	0.7593	0.6603	0.5871	0.5069	0.3585	0.2837	0.2058	0.1645	0.1284
2.0	0.9197	0.8528	0.7960	0.7047	0.6344	0.5544	0.3987	0.3168	0.2298	0.1832	0.1424
3.0	0.9219	0.8564	0.8006	0.7105	0.6406	0.5608	0.4044	0.3216	0.2334	0.1859	0.1445
4.0	0.9222	0.8569	0.8012	0.7113	0.6415	0.5617	0.4051	0.3222	0.2338	0.1863	0.1448
5.0	0.9222	0.8659	0.8013	0.7114	0.6416	0.5618	0.4052	0.3223	0.2339	0.1864	0.1448
∞	0.9222	0.8569	0.8012	0.7112	0.6414	0.5616	0.4050	0.3220	0.2335	0.1860	0.1444

Spatial problem. TBC $q_0 = $ const. Values of the factor of conjugation

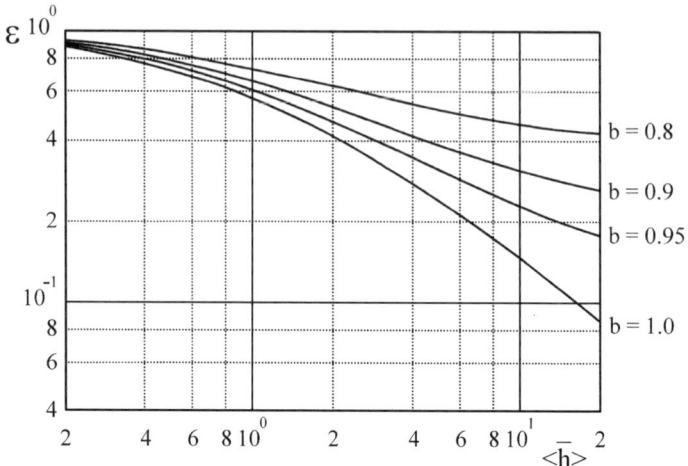

Fig. 3.9. Step pulsations. Spatial problem for a semi-infinite body. Values of the factor of conjugation

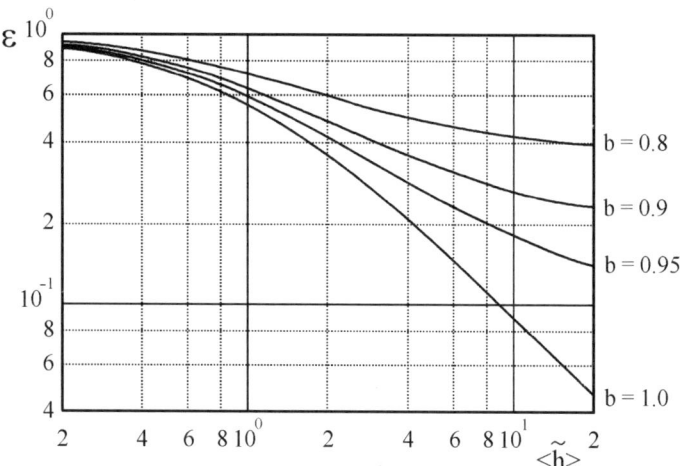

Fig. 3.10. Step pulsations. Time-dependent problem for a semi-infinite body. Values of the factor of conjugation

takes a self-identical form

$$\varepsilon = 1 - b^2 + \frac{8b^2}{\pi^2} f\left(\langle \bar{h} \rangle\right). \tag{3.70}$$

Substituting function $f(y) = 1/\left[y\left(y + \langle \bar{h} \rangle\right)\right]$ into (3.67) and taking its limit at $\langle \bar{h} \rangle \to \infty$, one can obtain the following expansion instead of (3.65)

$$\varepsilon = 1 - b^2 + \frac{4b^2}{\pi^2} \frac{\ln\left(\langle \bar{h} \rangle\right)}{\langle \bar{h} \rangle}. \tag{3.65a}$$

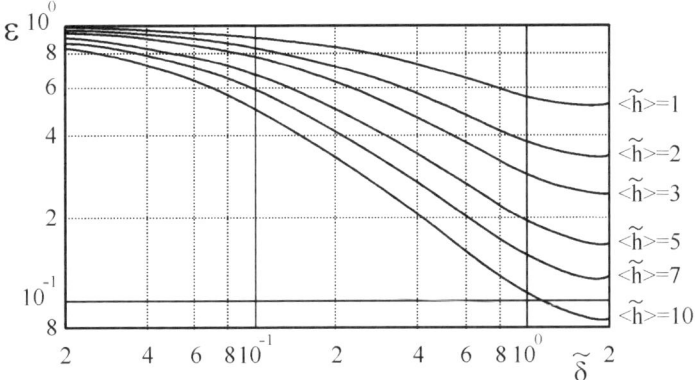

Fig. 3.11. Step pulsations. Time-dependent problem. TBC $\vartheta_0 = $ const. Values of the factor of conjugation

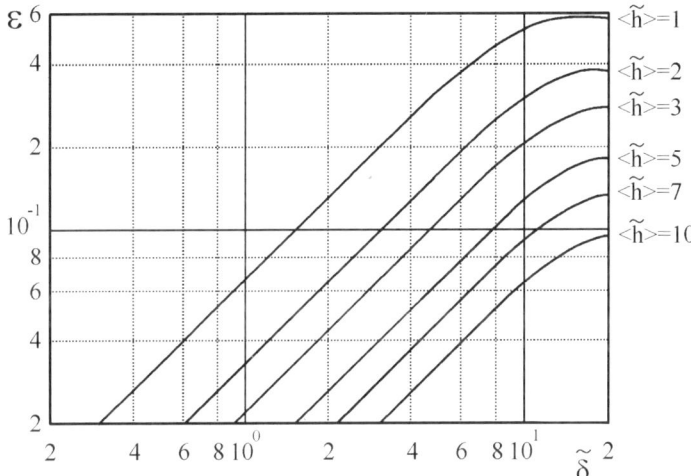

Fig. 3.12. Step pulsations. Time-dependent problem. TBC $q_0 = $ const. Values of the factor of conjugation

Thus, contrary to the time-dependent problem, the asymptotical solution for the FC for large Biot numbers in the case of spatial oscillations includes a logarithmic term.

Asymptotical solution at $\delta \to 0$ *for the TBC* $\vartheta_0 = $ const. This asymptotical solution can be written as

$$\varepsilon = 1 - b^2 \langle \bar{h} \rangle \bar{\delta}, \qquad (3.71)$$

or, in more obvious form,

$$\varepsilon = 1 - b^2 \frac{\langle h \rangle \delta}{k}. \qquad (3.72)$$

66   3 Solution of Characteristic Problems

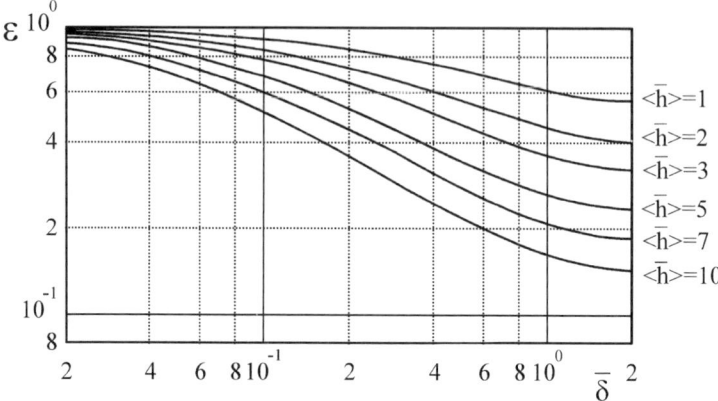

Fig. 3.13. Step pulsations. Spatial problem. TBC $\vartheta_0 = $ const. Values of the factor of conjugation

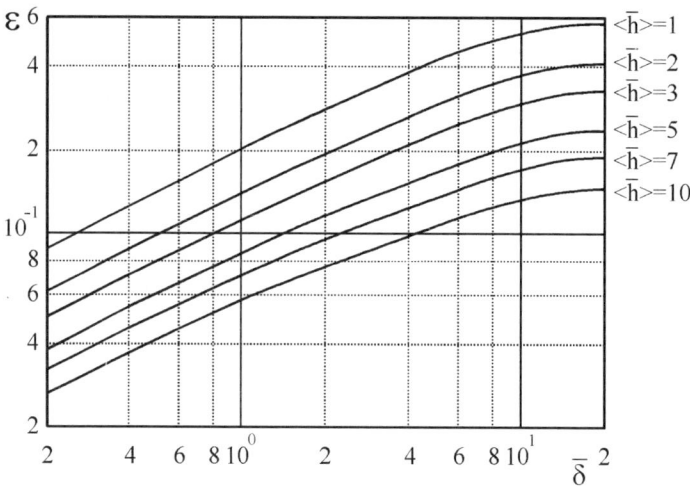

Fig. 3.14. Step pulsations. Spatial problem. TBC $q_0 = $ const. Values of the factor of conjugation

Asymptotical solution at $\delta \to 0$ *for the TBC* $q_0 = $ const. This asymptotical solution can be calculated with the help of a series expansion [8]

$$\tanh\left(\frac{\pi y}{2}\right) = \frac{4y}{\pi} \sum_{n=1}^{\infty} \frac{1}{(2n-1)^2 + y^2}. \qquad (3.73)$$

One can deduce from here that for a time-dependent problem

$$\varepsilon = 1 - b^2 + \frac{2b^2}{\pi} \frac{\tilde{\delta}}{\left\langle \tilde{h} \right\rangle} \qquad (3.74)$$

and
$$\varepsilon = 1 - b^2 + \frac{2b^2}{\pi}\sqrt{\frac{\delta}{\langle \bar{h} \rangle}} \qquad (3.75)$$

for a spatial problem, respectively.
Representation of the solution in a reduced form. It is interesting to note that solution (3.56) can be presented in the following reduced form
$$\varepsilon = \varepsilon_{\min} + (1 - \varepsilon_{\min})\,\varepsilon_*. \qquad (3.76)$$

Here
$$\varepsilon_* = 1 - \frac{8\langle \bar{h}\rangle}{\pi^2}\sum_{n=1}^{\infty}\frac{1}{(2n-1)^2}\frac{\langle \bar{h}\rangle + F_{2n-1}}{\left(\langle \bar{h}\rangle + F_{2n-1}\right)^2 + \Phi_{2n-1}^2} \qquad (3.77)$$

is the value of ε_1 at $b=1$; $\varepsilon_{\min} = 1 - b^2$ is the minimally possible value of the FC. The remarkable property of the solution for a symmetric step function mentioned above will be used below at an analysis of the corresponding asymmetrical case. For this reason, under the solution of a symmetric problem below we shall always mean expression (3.56) or equivalent system of (3.76), (3.77) [8].
Calculation of the sums of infinite series. While performing computations, the following procedure has been used for an approximate calculation of a sum of the series in solutions (3.56) and (3.59). Let us assume that it is necessary to calculate (from the formula (3.56)) the value ε_1 of the FC with a preset accuracy $\Delta(\varepsilon_1)$. Using a relation for the sum of a tabulated series
$$\sum_{n=1}^{\infty}\frac{1}{(2n-1)^2} = \frac{8}{\pi^2}, \qquad (3.78)$$

one can rewrite solution (3.56) in an equivalent form
$$\varepsilon_1 = 1 - b^2 + \frac{8b^2\langle \bar{h}\rangle}{\pi^2}\sum_{n=1}^{\infty}\frac{1}{(2n-1)^2}\frac{F_{2n-1}\left(\langle \bar{h}\rangle + F_{2n-1}\right) + \Phi_{2n-1}^2}{\left(\langle \bar{h}\rangle + F_{2n-1}\right)^2 + \Phi_{2n-1}^2}. \qquad (3.56a)$$

Since both series in expressions (3.56) and (3.56a) have a fixed sign, then the sequence of corresponding n-series (i.e., those terminated with a n-term) is the upper majorant for a hypothetical exact value ε_1 obtained at $K \to \infty$. Therefore, it is possible to estimate a maximal relative error of the approximate value ε_1 using a formula
$$\Delta(\varepsilon_1)_{\max} = \frac{|(\varepsilon_1)_n - (\varepsilon_1)_{n-1}|}{\min\{(\varepsilon_1)_n,(\varepsilon_1)_{n-1}\}}. \qquad (3.79)$$

Exactly the same reasoning can be carried out also for a calculation of the FC ε_2 with a preset accuracy $\Delta(\varepsilon_2)$ using (3.59). Comparison of the computed values $\varepsilon_1, \varepsilon_2$ among themselves gives the relative error of the final value of the FC.

Regular regime of periodic oscillations. Rather interesting is an asymptotical solution $\langle \tilde{h} \rangle \to \infty$ for a purely temporal step law of oscillations of the THTC. Assuming $b = 1$ in the series (3.64a) and considering approximately that $0.9679 \approx 1$, one can obtain: $\varepsilon = 1/\langle \tilde{h} \rangle$. In view of the expressions for the dimensionless ATHTC $\langle \tilde{h} \rangle = \langle h \rangle \sqrt{\alpha \tau_0}/k$ and for the FC $\varepsilon = h_m/\langle h \rangle$, one can derive a formula for the EHTC $h_m = \sqrt{kc\rho/\tau_0}$. Thus, at step temporal intermittency of a heat transfer period ($\psi = 1, h/\langle h \rangle = 2$) and adiabatic period ($\psi = -1, h/\langle h \rangle = 0$), in the limiting case of the infinitely large Biot numbers the EHTC ceases to depend on the ATHTC and is determined only by the factor of thermal activity of a solid body $\sqrt{kc\rho}$ and by the period of oscillations τ_0. This unexpected result evidently shows that the account of the thermal conjugation "fluid flow – body" can give not only quantitative corrections to the theoretical stationary heat transfer coefficient, but also result in qualitatively new results. The remarkable property of the solution found above can be called a regular regime of periodic oscillations (by analogy with a known regular regime of heat transfer [18]).

3.6 Comparative Analysis of the Conjugation Effects (Smooth and Step Oscillations)

One should point out at the beginning to this section that the symmetric step type of the periodic oscillations of the THTC investigated above is characterized by considerably stronger influence of a solid body on the EHTC in comparison with the smooth laws of oscillations considered in the preceding sections (harmonic and inverse harmonic ones). We intend to perform below a physical interpretation of the revealed tendencies of influence of different parameters in this problem for the case of temporal oscillations, which is the most important in engineering applications.

Semi-infinite body. Influence of the Biot number. If a wall thickness is much larger than the length of penetration of a thermal wave $\left(\delta >> \sqrt{\alpha \tau_0}, \tilde{\delta} >> 1\right)$, then the FC (for each value of the dimensionless fluctuation amplitude b) depends only on the Biot number $\langle \tilde{h} \rangle = \langle h \rangle \sqrt{\alpha \tau_0}/k$. It is obvious from Fig. 3.10 that the value of ε decreases together with the factor of thermal activity of a solid body $\sqrt{kc\rho}$. This effectively means that the distinction between the ATHTC and the EHTC increases, and thermal influence of a solid body increases as well. At the same value of the fluctuation amplitude b, the Biot number effect is pronounced stronger for the step function $\psi(t)$ (Fig. 3.10), than for the smooth functions $\psi(t)$, both harmonic (Fig. 3.3) and inverse harmonic one (Fig. 3.6). It is rather obvious from a physical reasoning: sharper oscillations of the heat transfer intensity result in a stronger interaction "body – fluid". An increase in the dimensionless amplitude of oscillations b (with other parameters being the same) results in a decrease in the FC for all the studied functions $\psi(t)$ (see Figs. 3.3, 3.6, and 3.10). In this case, an form of the dependence $\varepsilon(b)$ is essentially different for both symmetric (harmonic and step) and asymmetric (inverse harmonic) functions $\psi(t)$. In the first two cases, the form of the dependence $\psi(t)$ does not change qualitatively at $b \to 1$ (see Figs. 3.3 and 3.10). Respective subsiding dependence $\varepsilon\left(\left\langle \tilde{h} \right\rangle\right)$ changes only quantitatively, since it acquires a more pronounced steepness and finally falls down in a limit at $\left\langle \tilde{h} \right\rangle \to \infty$ to a level that is minimally possible for the given type of oscillations: $\varepsilon \to \varepsilon_{\min}$. In the second case at $b \to 1$, the function $\psi(t)$ undergoes a qualitative transformation and in a limit at $b = 1$ transfers in the Kroeneker delta-function. At increasing Biot numbers, the dependence of $\varepsilon\left(\left\langle \tilde{h} \right\rangle\right)$ falls down more and more abruptly from $\varepsilon = 1$ at $\left\langle \tilde{h} \right\rangle = 0$ to $\varepsilon = 0$ at $\left\langle \tilde{h} \right\rangle \to \infty$ (see Fig. 3.6).

Flat plate. Influence of a wall thickness. A dependence of the FC on a wall thickness starts to become perceptible when the latter becomes commensurable with the length of penetration of a thermal wave $\left(\delta \leq \sqrt{\alpha \tau_0}, \tilde{\delta} \leq 1\right)$. A particular character of this dependence (which amplifies with the reducing wall thickness) is determined by a type of the TBC. Thinner walls exhibit increased values of the FC at $\vartheta_0 = $ const and decreased FC at $q_0 = $ const (see Figs. 3.11 and 3.12). In a limit at $\delta \to 0$, we will have: $\varepsilon = 1$ at $\vartheta_0 = $ const (Fig. 3.11), $\varepsilon = \varepsilon_{\min}$ at $q_0 = $ const (Fig. 3.12). As one can see from Figs. 3.11 and 3.12, these tendencies of dependences $\varepsilon\left(\tilde{\delta}\right)$ computed for the step law of oscillations (at the largest possible amplitude $b = 1$) enforce at the increased Biot numbers.

3.7 Particular Exact Solution

As it was already emphasized in the above sections and chapters, the mainstream of a developed method will consist in the consecutive solution of the heat conduction equation with a periodic BC preset in a form of a specified harmonic, inverse harmonic or symmetric step function $\psi(\xi)$. The resulting analytical solutions are cumbersome, as they have a form of sums of infinite complex conjugate chain fractions and infinite functional series. Therefore, it is of a significant interest to obtain a simple exact solution, on the basis of which general properties and features of the developed analytical method can be investigated. Let us find periodic components of the temperatures and heat fluxes on a heat transfer surface as the first harmonic of a Fourier series [16]

$$\theta = R\cos(\xi), \tag{3.80}$$
$$\theta^\bullet = R\left[F\cos(\xi) - \Phi\sin(\xi)\right]. \tag{3.81}$$

As follows from (3.80) and (3.81), a fluctuation of the temperature gradient can be written down as

$$\theta^{\bullet} = R \cos(\xi + \xi_0), \tag{3.82}$$

where $\xi_0 = \arctan(F/\Phi)$. Substituting (3.80) into the BC (2.29), one can find the periodic component of the THTC

$$\psi = \frac{\varepsilon + R\left[\Phi \sin(\xi) - F \cos(\xi)\right]}{1 + R \cos(\xi)} - 1. \tag{3.83}$$

Let us use the designations $f = F/\langle \bar{h} \rangle$, with the FC being calculated from the ratio

$$\varepsilon = (1 + f)\sqrt{1 - R^2} - f. \tag{3.84}$$

Equation (3.84) has the following advantage: it includes only one generalized parameter of the thermal influence of a body f. However, this solution does not in principle agree with the physical model of the "hydrodynamically" determined heat transfer processes with periodic intensity. As follows from an analysis of (3.83), at increase in the parameter of the thermal influence of a body from zero (zero thermal conductivity of a body, maximal effect of the conjugation) to infinity (infinite thermal conductivity of a body, minimal effect of the conjugation), the function $\psi(\xi)$ evolves from the inverse harmonic (Fig. 3.4) up to the harmonic one (Fig. 3.1). In principle, such a behavior of the THTC does not agree with the basic concept of the method developed here. Indeed, function $\psi(\xi)$ is not anymore "hydrodynamically determined," but, on the contrary, artificially adjusts itself to the parameters of conjugation. On the other hand, the obtained simple solution is, apparently, a unique exact analytical solution of this problem. Therefore, it can be used as a test case for a validation and an estimation of other (more complex) solutions.

References

1. Zudin YB, Labuntsov DA (1978) Peculiarities of heat transfer at periodic asymmetrical regime. Works of Moscow Power Engineering Institute. Issue 377: 35–39 (in Russian).
2. Zudin YB (1980) Analysis of heat-transfer processes of periodic intensity. Dissertation. Moscow Power Engineering Institute (in Russian).
3. Labuntsov DA, Zudin YB (1984) Heat-Transfer Processes of Periodic Intensity Energoatomizdat, Moscow (in Russian).
4. Zudin YB (1991) A method of heat-exchange calculation in the presence of periodic intensity fluctuations. High Temp 29: 921–928.
5. Zudin YB (1994) Calculation of effect for supplying heat to the wall on the averaged heat exchange coefficient. Thermophys Aeromech 1: 117–119.
6. Zudin YB (1995) Design of the wall heat effect on averaged convective heat transfer in processes of heat exchange with periodic intensity. Appl Energy Russ J Fuel Power Heat Syst 33: 76–81.

7. Zudin YB (1996) Pulse law of true heat transfer coefficient pulsatinons. Appl Energy Russ J Fuel Power Heat Syst 34: 142–147.
8. Zudin YB (1996) Theory on heat-transfer processes of periodic intensity. Habilitation Moscow Power Engineering Institute (in Russian).
9. Zudin YB (1998) Effect of the thermophysical properties of the wall on the heat transfer coefficient. Therm Eng 45(3): 206–209.
10. Zudin YB (1998) Temperature waves on a wall surface. Russ Dokl Phys J Acad Sci 43(5): 313–314.
11. Zudin YB (1999) The effect of the method for supplying heat to the wall on the averaged heat-transfer coefficient in periodic rate heat-transfer prozesses. Therm Eng 46(3): 239–243.
12. Zudin YB (1999) Harmonic law of fluctuations of the true heat transfer coefficient. Thermophys Aeromech 6: 79–88.
13. Zudin YB (1999) Self-oscillating process of heat exchange with periodic intensity. J Eng Phys Thermophys 72: 635–641.
14. Zudin YB (2000) Processes of heat exchange with periodic intensity. Therm Eng 47(6): 124–128
15. Zudin YB (2000) Averaging of the heat-transfer coefficient in the processes of heat exchange with periodic intensity. J Eng Phys Thermophys 73: 643–647.
16. Stein EM, Shakarchi R (2003) Fourier Analysis: An Introduction. Princeton University Press, Princeton.
17. Tsoi PV, Tsoi VP (2002) A method of representing nonstationary temperature fields in the best approximations. High Temp 40: 456–468.
18. Baehr HD, Stephan J (1998) Heat and Mass Transfer. Springer, Berlin Heidelberg New York.
19. Dietz C, Henze M, Neumann SO, von Wolfersdorf J, Weigand B (2005) Numerical and experimental investigation of heat transfer and fluid flow around a vortex generator using explicit algebraic models for the turbulent heat flux. Proc. of the 17th Int. Symp. on Airbreathing Engines, September 4–9, Munich, Germany, Paper ISABE-2005-1197.
20. Khinchin AY (1997) Continued Fractions. Dover Publications, New York.
21. Sauer R, Szabo I (1969) Mathematische Hilfsmittel des Ingenieurs, Teil 1–4. Springer, Berlin.
22. Kantorovich LV, Krylov VI (1958) Approximate Methods of Higher Analysis. P. Noordho, Groningen, The Netherlands.
23. Abramovitz M, Stegun IA (1974) Handbook of Mathematical Functions with Formulas, Graphs, and Mathematical Tables. Dover Publications, New York.
24. Bronstein IN, Semendjajew KA, Musiol G, Mühlig H (2005) Taschenbuch der Mathematik. Verlag Harry Deutsch, Frankfurt.

4

Universal Algorithm of Computation of the Factor of Conjugation

4.1 Smooth Oscillations (Approximate Solutions)

The system of (3.7)–(3.9) presented in Chap. 3 allows in principle determining both all the eigenvalues A_n, A_n^*, as well as the factor of conjugation (FC), which is the key value of the analysis. However, at the same time the solutions obtained on its basis are very cumbersome (infinite complex conjugate chain fractions, infinite functional series). An advantage of these solutions (as well as of any analytical solution in general) consists in an opportunity of carrying out an asymptotic analysis and approximate estimations of tendencies, which can be exhibited by the dependence of the FC on the dimensionless parameters. However, a calculation of the FC for some particular values of the parameters requires carrying out numerical calculations. Such calculations have been performed in Chap. 3 for three characteristic laws of oscillations of the true heat transfer coefficient (THTC): harmonic, inverse harmonic, and stepwise. The tables and nomograms computed on their basis provide an opportunity for engineers and physicists to operate with concrete numerical values and also show a quantitative measure of influence of a solid body on the heat transfer characteristics.

However, based on such a bedrock, it is impossible to obtain a comprehensive solution of a problem of conjugate convective heat transfer. Certain questions still arise; to mention a few of them:

- What is the effect of conjugation for the functions $\psi(\xi)$ that differ from those mentioned as three basic ones?
- How will the FC react on an imposition of various disturbances (peaks of amplitude, modulation of a high-frequency component, etc.) on the functions $\psi(\xi)$?
- How will the character of conjugation change at spatial extension of the oscillations?
- Can the developed method allow a generalization for stochastic and non-periodic oscillations of thermohydraulic parameters?

To answer these (and possible other) questions, it is necessary to find out an effective approximate method for a solution of the problem of conjugate periodic heat transfer. The objective of the present chapter thus consists in a search for a universal algorithm of an approximate calculation of the FC. In the beginning, let us consider smooth functions $\psi(\xi)$ (harmonic and inverse harmonic).

Harmonic law of oscillations of the THTC. Let us consider the harmonic law of oscillations of the THTC determined by (3.10). Using the form of notation of the chain fractions through n-partial sums s_n, s_n^*, one can rewrite the exact solution (3.12) in the following form [1–3]

$$\left.\begin{array}{l} \varepsilon = 1 - \frac{b}{2}\left(\frac{1}{s_1} + \frac{1}{s_1^*}\right), \\ s_1 = c_1 - \frac{1}{s_2}, \quad s_1^* = c_1^* - \frac{1}{s_2^*}, \\ s_2 = c_2 - \frac{1}{s_3}, \quad s_2^* = c_2^* - \frac{1}{s_3^*}, \\ \vdots \\ s_n = c_n - \frac{1}{s_{n+1}}, \quad s_n^* = c_n^* - \frac{1}{s_{n+1}^*}. \end{array}\right\} \quad (4.1)$$

Here

$$c_n = \frac{2}{b}\left(1 + \frac{B_n}{\langle \bar{h} \rangle}\right), \quad c_n^* = \frac{2}{b}\left(1 + \frac{B_n^*}{\langle \bar{h} \rangle}\right),$$

with $n = 1, 2, 3, \ldots$.

Let us fulfill a procedure of a termination of an infinite chain fraction according to work [4]. For this purpose, let us also assume that all the eigenvalues are equal among themselves: $B_2 = B_3 = \ldots B_n \ldots = B_1$, $B_2^* = B_3^* = \ldots B_n^* \ldots = B_1^*$. Based on these assumptions, one can obtain the following approximate solution

$$\varepsilon = 1 - \left(b^2/2\right)\left(1/S + 1/S^*\right), \quad (4.2)$$

where

$$\left.\begin{array}{l} S = 1 + \frac{B_1}{\langle \bar{h} \rangle} + \left[\left(1 + \frac{B_1}{\langle \bar{h} \rangle}\right)^2 - b^2\right]^{1/2}, \\ S^* = 1 + \frac{B_1^*}{\langle \bar{h} \rangle} + \left[\left(1 + \frac{B_1^*}{\langle \bar{h} \rangle}\right)^2 - b^2\right]^{1/2}. \end{array}\right\} \quad (4.3)$$

As against to the computation of the sums of the infinite chain fractions [(4.1)], a calculation based on (4.2) does not cause any difficulty. For a spatial type of oscillations ($m = 0$) one can deduce: $B_1 = B_1^* = F_1$. In this case, (4.3) can be written in a real form

$$S = S^* = \left(1 + \frac{F_1}{\langle \bar{h} \rangle}\right) + \left[\left(1 + \frac{F_1}{\langle \bar{h} \rangle}\right)^2 - b^2\right]^{1/2}. \quad (4.4)$$

Here $F_1 = \coth\bar{\delta}$ for the TBC $\vartheta_0 = \text{const}$; $F_1 = \tanh\bar{\delta}$ for the TBC $q_0 = \text{const}$
One should also point out that an attempt to get rid of the complex conjugate values $B_1 = F_1 + i\Phi_1$, $B_1^* = F_1 - i\Phi_1$ under the radical in (4.3) and to write down a solution in a real form can finally result here in obtaining extremely cumbersome formulas.

Inverse harmonic law of oscillations of the THTC. Let us consider now the inverse harmonic law of oscillations of the THTC determined by (3.24). The reasoning similar to that presented above for the harmonic law gives, instead of (3.25), an approximate solution [5]

$$\varepsilon = \frac{\sqrt{1-b^2}}{1-(b^2/2)(1/S+1/S^*)}, \qquad (4.5)$$

where

$$\begin{aligned} S &= 1 + \frac{\langle\bar{h}\rangle\sqrt{1-b^2}}{B_1} + \left[\left(1 + \frac{\langle\bar{h}\rangle\sqrt{1-b^2}}{B_1}\right)^2 - b^2\right]^{1/2}, \\ S^* &= 1 + \frac{\langle\bar{h}\rangle\sqrt{1-b^2}}{B_1^*} + \left[\left(1 + \frac{\langle\bar{h}\rangle\sqrt{1-b^2}}{B_1^*}\right)^2 - b^2\right]^{1/2}. \end{aligned} \qquad (4.6)$$

For the spatial type of oscillations, one can obtain a real form of a notation of the solution

$$S = S^* = 1 + \frac{\langle\bar{h}\rangle\sqrt{1-b^2}}{F_1} + \left[\left(1 + \frac{\langle\bar{h}\rangle\sqrt{1-b^2}}{F_1}\right)^2 - b^2\right]^{1/2}. \qquad (4.7)$$

One should point out again that it is practically impossible also here to get rid of the complex conjugate values $B_1 = F_1 + i\Phi_1$, $B_1^* = F_1 - i\Phi_1$ under the radical in (4.6). The obtained approximate solutions for two smooth laws of the variation of the THTC mean actually a step forward in the development of the proposed method. The approximate solutions (4.2) and (4.3), and (4.5) and (4.6) are by far simpler than the initial exact solutions (3.12), (3.25). However, this progress relates only to two particular laws of the variation of the THTC, namely, harmonic and inverse harmonic. At the same time, an investigator is interested in solving a global problem, i.e., in gaining an opportunity to compute the FC for any type of periodic oscillations of the heat transfer intensity. For this purpose, it is necessary in ideal to create a universal algorithm for the calculation of the FC.

4.2 BC on a Heat Transfer Surface (Series Expansion in a Small Parameter)

The following step includes an asymptotic analysis of the boundary condition (BC) on an internal surface of a plate $X = \delta$. As it is obvious from both forms of its notation (expressions (2.29) and (2.30)), each of them contains one dimensionless parameter. Therefore, it is expedient to expand this BC into power series in a small parameter [6–8].

Small parameter $\langle \bar{h} \rangle$. Let us derive an expansion of the BC in the form of (2.29) in a small parameter $\langle h \rangle \ll 1$. A comparison of the coefficients at the terms with identical power exponents $\langle \bar{h} \rangle^n$ gives the following chain of consecutive approximations:

$$\left.\begin{array}{l} \theta = \theta_0 + \theta_1 \langle \bar{h} \rangle + \theta_2 \langle \bar{h} \rangle^2 + \cdots + \theta_n \langle \bar{h} \rangle^n + \cdots, \\ \theta^\bullet = \theta_0^\bullet + \theta_1^\bullet \langle \bar{h} \rangle + \theta_1^\bullet \langle \bar{h} \rangle^2 + \cdots + \theta_n^\bullet \langle \bar{h} \rangle^n + \cdots, \\ \varepsilon = \varepsilon_0 + \varepsilon_1 \langle \bar{h} \rangle + \varepsilon_2 \langle \bar{h} \rangle^2 + \cdots + \varepsilon_n \langle \bar{h} \rangle^n + \cdots, \\ n = 1, 2, 3, \ldots. \end{array}\right\} \quad (4.8)$$

$$\left.\begin{array}{l} (0)\ \theta_0 = \theta_0^\bullet = 0, \\ (1)\ \varepsilon_0 = 1, \quad \theta_1^\bullet = -\psi, \\ (2)\ \varepsilon_1 = \langle \psi\theta_1 \rangle, \quad \theta_2^\bullet = -\theta_1 + \psi\theta_1 - \langle \psi\theta_1 \rangle, \\ \quad\quad\quad \vdots \\ (n+1)\ \varepsilon_n = \langle \psi\theta_n \rangle, \quad \theta_{n+1}^\bullet = -\theta_n + \psi\theta_n - \langle \psi\theta_n \rangle. \\ n = 1, 2, 3, \ldots. \end{array}\right\} \quad (4.8\text{a})$$

For the harmonic type of oscillations of the THTC, the obtained recurrent formulas can be written out in an explicit form. We shall write down here only the first two terms of the expansion of the FC in the parameter $\langle \bar{h} \rangle \ll 1$

$$\varepsilon = 1 - \frac{b^2}{2} \frac{F_1}{F_1^2 + \Phi_1^2} \langle \bar{h} \rangle. \quad (4.9)$$

Small parameter $\langle \bar{h} \rangle^{-1}$. Let us obtain now an expansion of the BC in the form of (2.30) in the small parameter $\langle \bar{h} \rangle^{-1} \ll 1$. For this purpose, let us further denote: $c = \langle 1/(1+\psi) \rangle^{-1}$, $1 + \phi = c/(1+\psi)$. A chain of the consecutive approximations gives:

$$\left.\begin{array}{l} \theta = \theta_0 + \theta_1 \langle \bar{h} \rangle^{-1} + \theta_2 \langle \bar{h} \rangle^{-2} + \cdots + \theta_n \langle \bar{h} \rangle^{-n} + \cdots, \\ \theta^\bullet = \theta_0^\bullet + \theta_1^\bullet \langle \bar{h} \rangle^{-1} + \theta_1^\bullet \langle \bar{h} \rangle^{-2} + \cdots + \theta_n^\bullet \langle \bar{h} \rangle^{-n} + \cdots, \\ \varepsilon = \varepsilon_0 + \varepsilon_1 \langle \bar{h} \rangle^{-1} + \varepsilon_2 \langle \bar{h} \rangle^{-2} + \cdots + \varepsilon_n \langle \bar{h} \rangle^{-n} + \cdots, \\ n = 1, 2, 3, \ldots. \end{array}\right\} \quad (4.10)$$

$$\left.\begin{array}{l} (0)\ \varepsilon_0 = c, \quad \theta_0 = \phi, \\ (1)\ \varepsilon_1 = \langle \phi\theta_0^\bullet \rangle, \quad c\theta_1 = \phi\varepsilon_1 - \theta_0^\bullet - \phi\theta_0^\bullet + \langle \phi\theta_0^\bullet \rangle, \\ \quad\quad\quad \vdots \\ (n)\ \varepsilon_n = \langle \phi\theta_{n-1}^\bullet \rangle, \quad c\theta_n = \phi\varepsilon_n - \theta_{n-1}^\bullet - \phi\theta_{n-1}^\bullet + \langle \phi\theta_{n-1}^\bullet \rangle, \\ n = 1, 2, 3, \ldots. \end{array}\right\} \quad (4.10\text{a})$$

For the inverse harmonic type of oscillations of the THTC, the first two terms of the series expansion of the FC in the parameter $\langle \bar{h} \rangle^{-1} \ll 1$ look like

$$\varepsilon = \sqrt{1-b^2} + \frac{b^2}{2} F_1 \langle \bar{h} \rangle^{-1}. \tag{4.11}$$

Basically, there are no obstacles for a further escalation of the order of the approximation and a calculation of the next (quadratic etc.) terms in the corresponding equations. However, this way does not promise a big success. Indeed, like in the previous section, we can proceed ahead dealing only with smooth functions $\psi(\xi)$. For the functions $\psi(\xi)$ of any other kind, a calculation of coefficients of the corresponding power series becomes a serious problem (if it is possible at all). A use of the method of a small parameter for a manipulation with these functions is exactly so inefficient, like the formal notation of the general solution of this problem (written in the beginning of Chap. 3) is. One should not forget also the following engineering rule [10], which can be formulated approximately in such a way: the first term of any Taylor series bears more information, than the resting whole infinite series.

Smooth oscillations. Small parameter b. If the dimensionless amplitude of oscillations tends to zero ($b \to 0$), a distinction between both smooth types of oscillations of the THTC (harmonic and inverse harmonic) vanishes. Both functions $\psi(\xi)$ can be described with a single relation $\psi = (b/2)[\exp(i\xi) + \exp(i\xi)]$. Let us designate local values of the temperature and heat flux at $X = \delta$ through ϑ, q, their aveage values through $\langle \vartheta_\delta \rangle, \langle q_\delta \rangle$, and their fluctuation values through $\hat{\vartheta}_\delta, \hat{q}_\delta$. Let us also introduce nondimensionalized values: $\tilde{\vartheta}_\delta = \hat{\vartheta}_\delta / \langle \vartheta_\delta \rangle, \tilde{q}_\delta = \hat{q}_\delta / \langle q_\delta \rangle$. Then at $b \to 0$, one can obtain asymptotic relations

$$\tilde{\vartheta}_\delta = -\frac{b}{2}\left[\frac{\exp(i\xi)}{1+B/\langle \bar{h} \rangle} + \frac{\exp(-i\xi)}{1+B^*/\langle \bar{h} \rangle}\right], \tag{4.12}$$

$$\tilde{q}_\delta = \frac{b}{2}\left[\frac{(B/\langle \bar{h} \rangle)\exp(i\xi)}{1+B/\langle \bar{h} \rangle} + \frac{(B^*/\langle \bar{h} \rangle)\exp(-i\xi)}{1+B^*/\langle \bar{h} \rangle}\right]. \tag{4.13}$$

4.3 Derivation of a Computational Algorithm

Let us begin now dealing with a global problem of a derivation of a universal approximate algorithm for the calculation of the FC. Let us present expressions for the oscillations of temperatures and heat fluxes in an approximate form

$$\left.\begin{array}{l}\vartheta_0 = \text{const}: \tilde{\vartheta} \approx A\frac{\sinh(gx)}{\sinh(g\delta)}\cos(\xi), \quad \tilde{\vartheta}^\bullet \approx Ag\frac{\cosh(gx)}{\sinh(g\delta)}\cos(\xi) \\ q_0 = \text{const}: \tilde{\vartheta} \approx A\frac{\cosh(gx)}{\cosh(g\delta)}\cos(\xi), \quad \tilde{\vartheta}^\bullet \approx Ag\frac{\sinh(gx)}{\cosh(g\delta)}\cos(\xi)\end{array}\right\}. \tag{4.14}$$

4 Universal Algorithm of Computation of the Factor of Conjugation

From (4.14), one can obtain a linear correlation between oscillations $\tilde{\vartheta}, \tilde{\vartheta}^\bullet$

$$\tilde{\vartheta}^\bullet = H\tilde{\vartheta}, \tag{4.15}$$

where $H = g\coth(g\bar{\delta})$ for the TBC $\vartheta_0 = \text{const}$; $H = g\tanh(g\bar{\delta})$ for the TBC $q_0 = \text{const}$ Substituting (4.14) in the heat conduction equation (2.7), averaging it with respect to the coordinate of the progressive wave ξ and squaring the result, one can find the unknown coefficient g: $g = (1+m^2)^{1/4}$. A use of (4.15) in the BC (2.29) gives

$$(1+\psi)(1+\theta) \approx \varepsilon - \chi\theta. \tag{4.16}$$

Here $\chi = H/\langle \bar{h} \rangle$ is the generalized *parameter of the thermal effect* (PTE) of a solid body equal to

$$\left.\begin{aligned}\vartheta_0 = \text{const} : \chi &= \frac{(1+m^2)^{1/4}\coth(g\bar{\delta})}{\langle \bar{h} \rangle} \\ q_0 = \text{const} : \chi &= \frac{(1+m^2)^{1/4}\tanh(g\bar{\delta})}{\langle \bar{h} \rangle}\end{aligned}\right\}. \tag{4.17}$$

Let us express the fluctuation of the temperature from (4.17)

$$\theta \approx -1 + \frac{\varepsilon + \chi}{1+\psi+\chi}. \tag{4.18}$$

Averaging both parts of (4.18) over the period and using the natural condition of periodicity $\langle \theta \rangle \equiv 0$, one can obtain a quadrature

$$\frac{1}{\varepsilon + \chi} = \left\langle \frac{1}{1+\psi+\chi} \right\rangle. \tag{4.19}$$

An approximate algorithm for a calculation of the FC follows from here

$$\varepsilon = 2\pi \left(\int_0^{2\pi} \frac{d\xi}{1+\chi+\psi(\xi)} \right)^{-1} - \chi. \tag{4.20}$$

Thus, for any preset periodic function $\psi(\xi)$, the quadrature (4.20) determines the required FC. From the physical point of view within the framework of the approximate solution, it is accepted that between the oscillations of three values considered in a method $\tilde{\vartheta}_\delta, \tilde{q}_\delta, \psi$ there is no phase shift at variation of a progressive wave. The computational algorithm (4.20) plays an extremely important role in the development of the approximate theory of conjugate periodic heat transfer. As it was already mentioned above, the first stage of this theory consisted in a transition from the initial convective–conductive problem to a boundary problem for the heat conduction equation in a body. A physical basis for such method was the concept of a THTC. At the second stage, it was possible to achieve a radical simplification of the computational aspect of the developed method. From the mathematical point of view, (4.20) represents a functional dependence, i.e., a dependence of the function $\varepsilon(\chi)$ on the function $\psi(\xi)$. The physical aspect of the algorithm consists in the filtrational character of the dependence of the FC on the oscillations of the THTC. It means that possible deformations of the function $\psi(\xi)$ under integral will be in any case smoothed out to some extent at a transition to the final function $\varepsilon(\chi)$. The replacement of the formal construction of the general solution (Sect. 3.1) with a calculation of the quadrature (4.20) translates the problem into the domain of distinct physical and engineering applications. Our subsequent task will become now a substantiation of universality of (4.20). It will be shown below that at a transition from a boundary problem for the heat conduction equation in a body to the calculation of a quadrature, there is no perceptible loss of accuracy at determining of the key value of a problem, i.e., the FC.

Approximate solution for smooth oscillations of the THTC. Let us write down approximate analytical solutions for the smooth functions $\psi(\xi)$ following from the computational algorithm given by (4.20). As a result, one can obtain the following relation: for the harmonic law

$$\varepsilon = \sqrt{\varepsilon_{min}^2 + 2\chi + \chi^2} - \chi \qquad (4.21)$$

for the inverse harmonic law

$$\varepsilon = \frac{\chi}{\sqrt{1 + 2\chi/\varepsilon_{min} + \chi^2} - 1}. \qquad (4.22)$$

Here $\varepsilon_{min} = \sqrt{1 - b^2}$ is a minimally possible value of the FC identical for both smooth functions. A comparison with the results obtained above with the help of the method of a small parameter reveals that relations (4.21) and (4.22) are asymptotically exact. For the approximate solution (4.21), the limiting case at $\varepsilon \to 1$ can be described by a ratio

$$\varepsilon = 1 - \frac{b^2}{2} \frac{\langle \bar{h} \rangle}{(F_1^2 + \Phi_1^2)^{1/2}}, \qquad (4.23)$$

that coincides with the exact solution for the harmonic function (3.12) simultaneously for three (out of four possible) variants:

- A spatial problem for $\delta \to \infty$
- A spatial problem for $\delta \to 0$
- A time-dependent problem for $\delta \to 0$

For the fourth variant, a time-dependent problem for $\delta \to \infty$, there is a difference in the numerical coefficient. For the solution (4.22), the limiting case $\varepsilon \to \varepsilon_{min}$ is described by the following relations: the time-dependent problem for semi-infinite bodies

$$\varepsilon = \sqrt{1 - b^2} + \frac{b^2}{2} \frac{1}{\langle \tilde{h} \rangle}, \qquad (4.24)$$

the spatial problem for semi-infinite bodies

$$\varepsilon = \sqrt{1-b^2} + \frac{b^2}{2}\frac{1}{\langle \tilde{h} \rangle}, \qquad (4.25)$$

the time-dependent problem for the TBC $q_0 = \text{const}$

$$\varepsilon = \sqrt{1-b^2} + \frac{b^2}{2}\frac{H}{\langle \tilde{h} \rangle}, \qquad (4.26)$$

the spatial problem for the TBC $q_0 = \text{const}$

$$\varepsilon = \sqrt{1-b^2} + \frac{b^2}{2}\frac{H}{\langle \tilde{h} \rangle}. \qquad (4.27)$$

4.4 Phase Shift Between Oscillations

Equation (4.20) can be considered as the first iteration in the procedure of an approximate calculation of the FC. Therefore, for a validation of its accuracy, it is necessary to find out a next-order approximation at the expense of an introduction of certain corrections in the computational algorithm. As it was already mentioned, the derivation of (4.20) was based on an assumption concerning a synchronism of oscillations of the values $\hat{\vartheta}, \hat{q}, \hat{h}$. Therefore, it is deemed natural in the second-order approximation to take into account phase shifts between oscillations of these parameters. Let us assume that the resulted fluctuation in the temperature of the heat transfer surface, $\tilde{\vartheta}_\delta = \hat{\vartheta}_\delta/\langle \vartheta_\delta \rangle, \tilde{q}_\delta = \hat{q}_\delta/\langle q_\delta \rangle$, can be written down as (4.18) with allowance for the phase shift ξ_ϑ in relation to the base oscillations of $\psi(\xi)$:

$$\tilde{\vartheta}_\delta = -1 + \frac{\varepsilon + \chi}{1 + \psi(\xi + \xi_\vartheta) + \chi}. \qquad (4.28)$$

Then the resulting fluctuation of the heat flux

$$\tilde{q}_\delta \equiv \frac{\hat{q}_\delta}{\langle q_\delta \rangle} = -\frac{\theta^\bullet}{\langle \tilde{h} \rangle \varepsilon}, \qquad (4.29)$$

in view of the phase shift ξ_q, will look like

$$\tilde{q}_\delta = \frac{\chi}{\varepsilon}\left[1 - \frac{\varepsilon + \chi}{1 + \psi(\xi + \xi_q) + \chi}\right]. \qquad (4.30)$$

4.4 Phase Shift Between Oscillations

Let us rewrite the two equivalent forms of the BC (2.29) and (2.30) in new notations

$$\varepsilon = 1 + \left\langle \psi \tilde{\vartheta}_\delta \right\rangle, \tag{4.31}$$

$$\varepsilon^{-1} = \left\langle \frac{1}{1+\psi} \right\rangle + \left\langle \frac{\tilde{q}_\delta}{1+\psi} \right\rangle. \tag{4.32}$$

At known values of the phase shifts ξ_ϑ, ξ_q, a substitution of (4.28) and (4.30) into (4.31) and (4.32) gives new values of the FC. Thus, in doing so one can obtain an approximate solution of a higher order in comparison with (4.20). It, in turn, will enable a validation and an improvement of the above-mentioned algorithm of the calculation of the FC. Knowledge of the phase shifts between oscillations of parameters allows analyzing directly oscillations of temperatures and heat fluxes in a body. For this purpose, it is necessary to substitute into expressions (4.28) and (4.30) the values of the FC computed according to the algorithm given by (4.20).

Harmonic law of oscillations of the THTC. For the harmonic law of oscillations of the THTC $\psi = b \cos \xi$, a determination of the phase shift can be carried out with the help of a series expansion in a small parameter $\langle \bar{h} \rangle \ll 1$. It follows from this method that the first approximation of the fluctuation of the temperature gradient looks like a cosine function: $\theta_1^\bullet = -\psi$. Let us also find out approximate relations for the periodic components of the temperatures and heat fluxes on a heat transfer surface [(3.44) and (3.45)] in a form of the first harmonics

$$\theta = R \cos \xi - I \sin \xi, \quad \theta^\bullet = (FR - \Phi I) \cos \xi - (FI + \Phi R) \sin \xi. \tag{4.33}$$

Then, a substitution of this result into the BC (4.31) gives

$$(1 + b \cos \xi)(1 + R \cos \xi - I \sin \xi)$$
$$= \varepsilon + (\varphi I - fR) \cos \xi + (fI + \varphi R) \sin \xi. \tag{4.34}$$

Here $f = F/\langle \bar{h} \rangle$, $\varphi = \Phi/\langle \bar{h} \rangle$. Averaging of both parts of (4.34) yields

$$1 + bR/2 = \varepsilon. \tag{4.35}$$

In turn, multiplying both parts of (4.34) by $\sin \xi$, one can obtain

$$R = -\frac{1+f}{\varphi} I. \tag{4.36}$$

After subsequent simple transformations, it follows from (4.28) and (4.30)

$$\theta \sim -\cos(\xi + \xi_\vartheta), \quad \theta^\bullet \sim \cos(\xi + \xi_q). \tag{4.37}$$

Here ξ_ϑ, ξ_q are the phase shifts determined by the relations

$$\xi_\vartheta = -\arctan\left(\frac{\varphi}{1+f}\right), \quad \xi_q = \arctan\left(\frac{\varphi}{\varphi^2 + f^2 + f}\right). \tag{4.38}$$

Thus, a fluctuation of the temperature θ is late in relation to the basic function ψ, and the fluctuation of a gradient of the temperature $\theta^\bullet$, on the contrary, outstrips the value of ψ. One should point out that the relations for $\theta, \theta^\bullet$ are written down to within a constant multiplying factor, since, at a determination of a phase, amplitudes of oscillations are insignificant. The negative sign in the expression for the fluctuation of the temperature θ is physically natural: for the considered case of cooling of a body, an increase in the heat transfer intensity leads to a reduction of the wall temperature.

Inverse harmonic law of oscillations of the THTC. Let us consider now oscillations of the THTC having a form of an inverse harmonic function:

$$\psi = \frac{\sqrt{1-b^2}}{1+b\cos\xi} - 1. \tag{4.39}$$

A determination of the phase shift can be carried out here with a help of the series expansion in a small parameter $\langle \bar{h} \rangle^{-1} \ll 1$. It follows further that, in the zeroth approximation, the temperature fluctuation looks like a cosine function $\theta_0 = \phi$. Then, a substitution into the BC (4.32) gives

$$\sqrt{1-b^2}\,(1+R\cos\xi - I\sin\xi)$$
$$= (1+b\cos\xi)\left[\varepsilon + (\varphi I - fR)\cos\xi + (fI + \varphi R)\sin\xi\right]. \tag{4.40}$$

Averaging of both parts of (4.40) yields

$$\sqrt{1-b^2} = \varepsilon + \frac{b}{2}(\varphi I - fR). \tag{4.41}$$

In turn, multiplying both parts of (4.40) by $\sin\xi$ results in

$$R = -\frac{f+\sqrt{1-b^2}}{\varphi}I. \tag{4.42}$$

After rather simple transformations, it follows from (4.41) and (4.42):

$$\xi_\vartheta = -\arctan\left(\frac{\varphi}{\sqrt{1-b^2}+f}\right), \quad \xi_q = \arctan\left(\frac{\sqrt{1-b^2}\varphi}{\varphi^2 + f^2 + f\sqrt{1-b^2}}\right). \tag{4.43}$$

As further calculations show, the first- and the second-order approximations of the solutions for the FC of the smooth functions $\psi(\xi)$ obtained on the basis of algorithm (4.20) coincide practically completely among themselves. It is also important to note the following circumstance. Approximate solutions of the first order have very simple forms of (4.21) and (4.22). At the same time, approximate solutions of the second order cannot be any more presented in an analytical form, and can be further obtained by a numerical computation of the corresponding quadratures. It confirms a notice about the priority importance of analytical solutions made at the end of Chap. 1. An analysis of the computational algorithm

4.5 Method of a Small Parameter

It has been convincingly shown above that the computational algorithm (4.20) is an effective tool for obtaining simple analytical solutions for the FC at a preset law of oscillations of the THTC. Let us show now that it is possible on its basis to carry out also an analysis of a behavior of the FC in a general form, i.e., for any periodic function $\psi(\xi)$. For this purpose, let us carry out an asymptotic investigation of the quadrature (4.19) [10].

Asymptotical solution at $\psi \to 0$. Denoting $c = \psi/(1+\chi)$, $Y = (1+\chi)/(\varepsilon + \chi)$, one can rewrite the quadrature (4.19) as

$$Y = \left\langle \frac{1}{1+c} \right\rangle. \tag{4.44}$$

Expanding the integrand expression in (4.44) in a power series at $c \to 0$ and swapping the operations of division and integration [10], one can obtain

$$Y = 1 + \langle c^2 \rangle + \langle c^4 \rangle + \cdots + \langle c^{2n} \rangle + \cdots, \quad n = 1, 2, 3, \ldots. \tag{4.45}$$

For the harmonic function $\psi(\xi)$, the expansion of (4.45) can be written out in an explicit form

$$y = 1 + \frac{1}{2}\eta + \frac{3}{8}\eta^2 + \frac{5}{16}\eta^3 + \frac{35}{128}\eta^4 + \ldots, \quad \eta = \left(\frac{b}{1+\chi}\right)^2. \tag{4.46}$$

Asymptotical solution at $\chi^{-1} \to 0$. An expansion of both parts of (4.19) in a series in a small parameter $\kappa = \chi^{-1}$ results in a power series of the following form

$$1 - \varepsilon\kappa + (\varepsilon\kappa)^2 - (\varepsilon\kappa)^3 + \cdots + (-1)^n (\varepsilon\kappa)^n + \cdots$$
$$= 1 - \langle 1 + \psi \rangle \kappa + \left\langle (1+\psi)^2 \right\rangle \kappa^2 - \left\langle (1+\psi)^3 \right\rangle \kappa^3$$
$$+ \cdots + (-1)^n \left\langle (1+\psi)^n \right\rangle \kappa^n + \cdots, \quad n = 1, 2, 3, \ldots. \tag{4.47}$$

As one can see from expansion (4.47), consecutive termination of the series at each number n results, at an increase of the latter, in obtaining a corresponding algebraic equation for the FC with a respectively growing order. Therefore, apparent simplicity of expansion (4.47) is, in fact, deceiving and this expansion can be actually realized only to within a linear term

$$\varepsilon = 1 - \langle \psi^2 \rangle \chi^{-1}. \tag{4.48}$$

One can further obtain from (4.48) an asymptotic relation for the harmonic function $\psi(\xi)$

$$\varepsilon = 1 - \frac{b^2}{2}\chi^{-1}, \tag{4.49}$$

and for the inverse harmonic function $\psi(\xi)$

$$\varepsilon = 1 - \frac{1 - \sqrt{1-b^2}}{\sqrt{1-b^2}}\chi^{-1}. \tag{4.50}$$

Expansions (4.49) and (4.50) entirely coincide with corresponding asymptotical solutions following from the approximate solutions (4.21) and (4.22). Indeed, one could expect obtaining a different result: the difference in the solution procedures here consists in a simple rearrangement of the operations of the series expansion and integration. Therefore, relations (4.49) and (4.50) play here a role of a kind of an original validation of the computational algorithm (4.19).

Asymptotical solution at $\chi \to 0$. A series expansion of the integrand expression in the quadrature (4.19) in a small parameter χ results in an infinite series of the following kind

$$\frac{\chi}{\varepsilon + \chi} = \left\langle \frac{1}{1+\psi} \right\rangle \chi - \left\langle \frac{1}{(1+\psi)^2} \right\rangle \chi^2 + \left\langle \frac{1}{(1+\psi)^3} \right\rangle \chi^3$$
$$- \cdots + (-1)^{n+1} \left\langle \frac{1}{(1+\psi)^n} \right\rangle \chi^n + \cdots, \quad n = 1, 2, 3, \ldots. \tag{4.51}$$

Here, as against to the previous case, a determination of the FC can be formally carried out down to any arbitrary value n. However, in this case, certain technical difficulties of another kind arise that are connected to a calculation of the integrals in the right-hand side of (4.51). Therefore, for real applications, expansion (4.51) can be also realized only to within a linear term

$$\varepsilon = \frac{1}{G_1} + \left(\frac{G_2}{G_1^2} - 1\right)\chi. \tag{4.52}$$

Here $G_1 = \langle 1/(1+\psi) \rangle$, $G_2 = \langle 1/(1+\psi)^2 \rangle$. One can obtain the following asymptotic expressions from (4.52): for the harmonic function $\psi(\xi)$

$$\varepsilon = \sqrt{1-b^2} + \frac{1 - \sqrt{1-b^2}}{\sqrt{1-b^2}}\chi, \tag{4.53}$$

for the inverse harmonic function $\psi(\xi)$

$$\varepsilon = \sqrt{1-b^2} + \frac{1}{2}b^2\chi. \tag{4.54}$$

Like in the above considered cases, expansions (4.53) and (4.54) also identically coincide with corresponding asymptotical solutions following from the approximate expressions (4.21) and (4.22). Thus, an application of the method of a small parameter within the framework of the approximate solution [(4.19) or (4.20)] gives practically the same effect, as the direct use of the approximate solution in the exact BC [(4.31) and (4.32)]. It may be further concluded from this fact that:

- There are no basic problems for a calculation of the second-order and subsequent higher-order terms of the series.
- The most preferable (and really possible) is a use of only two smooth functions $\psi(\xi)$.
- In the reality, one should be content only with the first (linear) terms of the power series.

However, one can notice some progress here that consists in the fact that each of the two expansions [(4.49) and (4.50) at $\chi^{-1} \to 0$; and (4.53) and (4.54) at $\chi \to 0$] are now equally suitable for both specified functions. It should be reminded also that each of the exact expansions allowed earlier an analytical representation only for one of the functions $\psi(\xi)$: (4.49) for the harmonic and (4.50) for the inverse harmonic law.

4.6 Application of the Algorithm for an Arbitrary Law of Oscillations

The approximate algorithm of a calculation of the FC (4.20) opens wide opportunities for investigations of arbitrary periodic functions $\psi(\xi)$. Its advantage consists also in the fact that the effects of the two major determining parameters, the Biot numbers and the wall dimensionless thickness, are concentrated in the value of the PTE [(4.17)]. As a result, the whole cumulative influence of the thermal effect of a solid body on the heat transfer characteristics is described by a single dependence $\varepsilon(\chi)$, which is universal for each preset function $\psi(\xi)$. Thus, an analysis of the thermal effect of a solid body in a problem of periodic heat transfer becomes a quite solvable problem. Let us illustrate an opportunity of an application of the method presented above at an example of nucleate boiling in a free infinite volume.

- The value of the averaged THTC $\langle h \rangle$ is borrowed from a corresponding stationary nonconjugate theory of the considered process and, consequently, it is known beforehand. It is possible, for example, to use here the known theory of nucleate boiling proposed by Labuntsov [11].
- The parameter $m = Z_0^2/(\alpha \tau_0)$ (inverse Fourier number) is calculated based on the thermal diffusivity of a body a, a distance between the centers of boiling Z_0 and the period of a life cycle of a particular steam bubble τ_0.
- Knowing spatial scale of oscillations Z_0, one determines the Biot number: $\langle \bar{h} \rangle = \langle h \rangle Z_0/k$ and the dimensionless wall thickness $\bar{\delta} = \delta/Z_0$.

- Conditions of external heat transfer are known and realized as a corresponding TBC ($\vartheta_0 = $ const, $q_0 = $ const).
- Some uncertainty will consist in the definition of a type of the periodic function $\psi(\xi)$ describing the mechanism of thermohydraulic oscillations. However, this problem is not of a fundamental, but rather of a physical nature and can be solved at the level of modeling of the boiling process [11, 12].

Characteristic examples describing a use of the approximate algorithm (4.20) are considered below.

Asymmetric smooth oscillations. An inverse harmonic function,

$$\frac{h}{\langle h \rangle} = 1 + \psi = \frac{\sqrt{1-b^2}}{1 + b\cos(\xi)}, \qquad (4.55)$$

describes oscillations with the limiting values of the amplitude equal to

$$1 + \psi_{\min} = \sqrt{\frac{1-b}{1+b}}, \quad 1 + \psi_{\max} = \sqrt{\frac{1+b}{1-b}}. \qquad (4.56)$$

Let us preset an asymmetric function $\psi(\xi)$ from a condition that the minimal heat transfer intensity over a period is equal to zero

$$\frac{h}{\langle h \rangle} = 1 + \psi = \frac{\sqrt{1-b}}{\sqrt{1+b} - \sqrt{1-b}} \left(\frac{1+b}{1+b\cos(\xi)} - 1 \right). \qquad (4.57)$$

In this case, the maximal value of the amplitude over a period is

$$1 + \psi_{\max} = \frac{2b}{\sqrt{1-b}\left(\sqrt{1+b} - \sqrt{1-b}\right)}. \qquad (4.58)$$

It is obvious from Fig. 4.1 that a transition from the usual function $\psi(\xi)$ to the asymmetric one results in an extension of its amplitude. Corresponding dependence $\varepsilon(\chi)$ in the latter case (which mathematical formulas are not presented here in view of its very cumbersome form) is much steeper. In the asymptotical case of the limiting thermal effect of a solid body, one can obtain here: $\chi \to 0, \varepsilon \to 0$. It should be reminded that for the initial function $\psi(\xi)$ determined by (4.22), the value of the FC in this asymptotical solution tends not to zero, but to some minimal value determined by the amplitude of oscillations: $\chi \to 0, \varepsilon \to \sqrt{1-b^2}$. It is rather interesting that the pointed out distinctions at use of these two functions $\psi(\xi)$ disappear at limiting transition $b \to 1$. Here both dependences $\varepsilon(\chi)$ have the same delta-like asymptotical form.

Saw-tooth form of oscillations. Saw-tooth oscillations of the THTC (Fig. 4.2) are described by means of a ratio

$$\frac{h}{\langle h \rangle} = 1 + \psi = \frac{b\{\exp[-b\xi/(2\pi)] - \exp(-b)\}}{1 - (1+b)\exp(-b)}. \qquad (4.59)$$

At $b \to 0$, the ratio (4.59) describes symmetric saw-tooth function

4.6 Application of the Algorithm for an Arbitrary Law of Oscillations

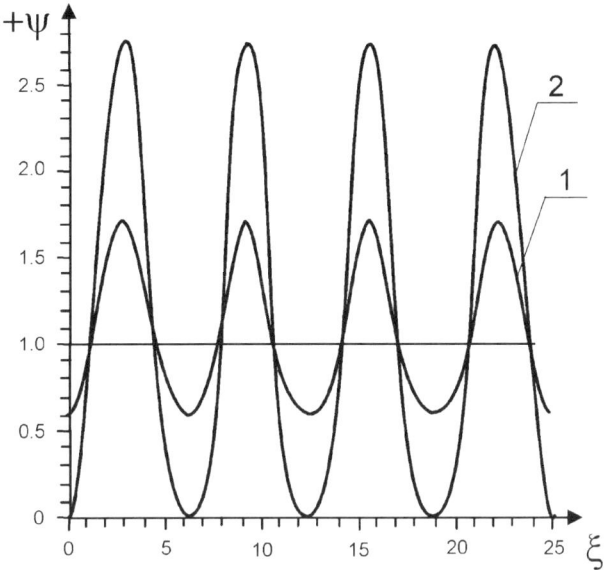

Fig. 4.1. Inverse harmonic law of pulsations of the THTC: (**a**) symmetric (1, (4.55)) and asymmetric (2, (4.57)) functions $\psi(\xi)$

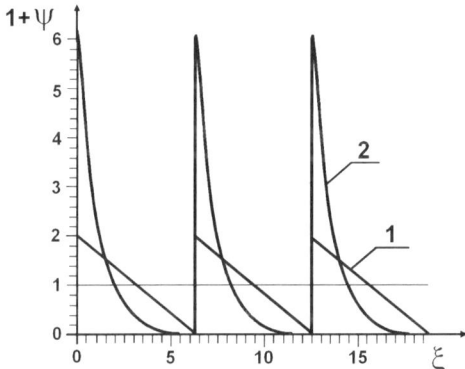

Fig. 4.2. Saw-tooth pulsations of the THTC (1, (4.60); 2, (4.59))

$$1 + \psi = 2 - \frac{\xi}{2\pi}. \tag{4.60}$$

At $b \to \infty$, the dependence of $h/\langle h \rangle$ acquires a characteristic delta-like form

$$1 + \psi \to 1 - \exp\left[-b\left(1 - \frac{\xi}{2\pi}\right)\right]. \tag{4.61}$$

The above-mentioned examples are the brightest illustrations of an application of the approximate algorithm (4.20). The list of possible periodic functions

$\psi(\xi)$ can be continued, since with the help of the universal approximate algorithm developed above it was possible to find out a way how to successfully bypass significant mathematical technical difficulties encountered in the previous chapters.

Inverse saw-tooth oscillations. As it was mentioned in the survey part of Chap. 1, at gravitational flow of a liquid film on a vertical surface, a regime can take place where discrete liquid volumes roll down over a surface of a thin liquid film of a practically constant thickness. The variation of a film thickness with the coordinate of a progressive wave has approximately a saw-tooth form

$$l = 2\langle l \rangle \left(1 - \frac{b\xi}{2\pi}\right). \tag{4.62}$$

The value of the THTC in this case can be quite precisely described by a dependence [13, 14] $h = k_f/\delta_f$, i.e., it has an inverse saw-tooth form. It follows from here that

$$1 + \psi = -\frac{1}{b\ln(1-b)(1-b\xi)}. \tag{4.63}$$

Two-dimensional harmonic oscillations (spatial problem). An interesting example of efficiency of the algorithm (4.20) is represented by a case with two-dimensional spatial oscillations of heat transfer intensity

$$\psi = b\cos(z)\cos(y). \tag{4.64}$$

The stationary three-dimensional heat conduction equation for a considered case looks like

$$\frac{\partial^2 \vartheta}{\partial X^2} + \frac{\partial^2 \vartheta}{\partial Z^2} + \frac{\partial^2 \vartheta}{\partial Y^2} = 0. \tag{4.65}$$

Owing to linearity of (4.65), its solution can be presented as a superposition of the stationary part $\langle \vartheta \rangle (X)$ satisfying the equation

$$\frac{\partial^2 \langle \vartheta \rangle}{\partial X^2} + \frac{\partial^2 \langle \vartheta \rangle}{\partial Z^2} + \frac{\partial^2 \langle \vartheta \rangle}{\partial Y^2} = 0 \tag{4.66}$$

and the fluctuation additive $\tilde{\vartheta}(X, Z, Y)$ described by the Laplace's equation

$$\frac{\partial^2 \tilde{\vartheta}}{\partial X^2} + \frac{\partial^2 \tilde{\vartheta}}{\partial Z^2} + \frac{\partial^2 \tilde{\vartheta}}{\partial Y^2} = 0. \tag{4.67}$$

The solution of (4.67) in its general form is represented by a double Fourier series [15, 16]. The algorithm (4.20) for the considered case can be written as

$$\varepsilon = (2\pi)^2 \left[\left(\int_0^{2\pi}\int_0^{2\pi} \frac{dz}{1+\chi+\psi(z,y)}\right)dy\right]^{-1} - \chi. \tag{4.68}$$

4.6 Application of the Algorithm for an Arbitrary Law of Oscillations

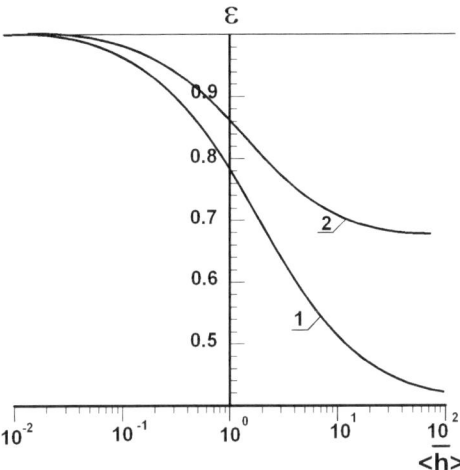

Fig. 4.3. One-dimensional (1, (4.21)) and two-dimensional (2, (4.69)) pulsations of the THTC (spatial problem, $b = 0.9$). Values of the factor of conjugation

For the function $\psi(z, y)$ defined by the ratio (4.64), a calculation of the quadrature (4.68) gives

$$\varepsilon = \frac{\pi}{2} \frac{1 + \chi}{E\left[b/(1 + \chi)\right]} - \chi, \qquad (4.69)$$

where $E(x)$ is an elliptic integral of the first kind [17]. As one can see from Fig. 4.3, the thermal effect of a body for the case of two-dimensional periodicity, at all the other conditions being equal, will be weaker, than that for the one-dimensional harmonic function (solution (4.21)). A minimal value of the FC for a two-dimensional case is

$$\varepsilon_{\min} = (2\pi)^2 \left[\left(\int_0^{2\pi} \int_0^{2\pi} \frac{dz}{1 + b \cos(z) \cos(y)} \right) dy \right]^{-1} = \frac{\pi}{2} \frac{1}{E(b)}. \qquad (4.70)$$

Dependences of the minimal value of the FC on the amplitude of spatial oscillations of the THTC for the one-dimensional and two-dimensional cases are shown in Fig. 4.4. As obviously follows from this figure, the dependence $\varepsilon_{\min}(b)$ for the latter case is higher, than for the former. From the physical point of view, it means weakening of the thermal effect of a body at spatially extended oscillations of the heat transfer intensity. In other words, the use of the computational algorithm allowed drawing a conclusion (in fact, nonevident beforehand) that surface temperature nonuniformity for the two-dimensional case is less pronounced than for the one-dimensional one. It is also possible to point out a certain analogy to the theory of turbulence. As it is known, in accordance with the theorem of Dwyer [18, 19], the two-dimensional perturbations imposed on the laminar fluid flow result in a stronger development of hydrodynamic

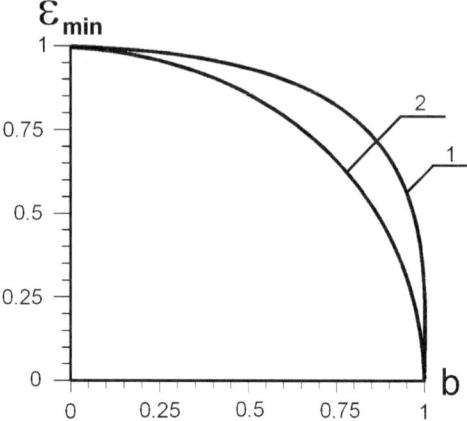

Fig. 4.4. Minimal value of the FC: two-dimensional (1, (4.69)) and one-dimensional (2, (4.21)) pulsations of the THTC

instability, than it happens in the cases of three-dimensional instabilities. In our case, it is possible to treat this situation in such a simplified way: a presence of the second spatial coordinate along a surface "body – fluid" provides a solid body with more opportunities for smoothing temperature nonuniformities in comparison with a one-dimensional case. Accordingly, the thermal effect on the fluid flow weakens and, as a consequence, a degree of conjugation of the convective–conductive heat transfer also decreases.

Two-dimensional inverse harmonic oscillations. Similarly to the previous case, one can also consider a two-dimensional inverse harmonic function

$$1 + \psi = \frac{\varepsilon_{\min}}{1 + b\cos(z)\cos(y)}. \tag{4.71}$$

A minimal value of the FC can be also calculated by consecutive averaging of (4.71) with respect to coordinates z, y. Thus, identically to the considered above one-dimensional case, values $\varepsilon_{\min}$ for both smooth functions coincide for a two-dimensional case. One should also point out that in the considered case the effect of the spatial extension of oscillations also takes place.

Oscillations of a standing wave type. Imposing of two progressive waves of an equal amplitude with opposite phase speeds result in a formation of a standing wave of the following kind

$$\psi = b\cos(t)\cos(z). \tag{4.72}$$

In this case, the heat conduction equation for the fluctuation component of a temperature field cannot be simplified to the form of (2.7) and will depend on three variables t, x, z:

$$m\frac{\partial \tilde{\vartheta}}{\partial t} - \frac{\partial^2 \tilde{\vartheta}}{\partial z^2} = \frac{\partial^2 \tilde{\vartheta}}{\partial x^2}. \tag{4.73}$$

An application of standard methods of separation of variables for a solution of (4.73) results in a triple Fourier series [15, 16] that are too cumbersome. Algorithm (4.20) for the considered case will be written down as (4.68) with the only distinction that an integration with respect to y is replaced with integration over t. In this case, the form of the solution is identical and looks as (4.70). However, functions of a body thickness include an additional parameter $m = Z_0^2/(\alpha\tau_0)$, or, in other words, the inverse Fourier number (see Appendix B). The limiting case of $m \to 0$ corresponds physically to an unlimited extension of the time period: $\tau_0 \to \infty$. In this case, the standing wave of oscillations is frozen in a form of a two-dimensional spatial function, and the solution passes into (4.70). As it was already discussed, the thermal effect of a body is weaker here, than for the one-dimensional harmonic law of oscillations. The limiting case of $m \to \infty$ is realized at an infinite speed of the propagation of oscillations on a surface of a body that corresponds to an unlimited extension of the spatial period of oscillations $Z_0 \to \infty$. So, the qualitative tendencies revealed earlier for spatiotemporal oscillations of the heat transfer intensity hold also for the case of a standing wave. In the quantitative sense, the thermal effect in the latter case is less, than in the former case. This interesting conclusion has a simple physical explanation. Indeed, at imposing of two opposite-directed progressive waves, fixed nodes are formed on a heat transfer surface, at which the amplitude of oscillations of the THTC is equal to zero. At approaching from the nodes toward the center of a periodic cell, the amplitude will monotonically increase up to its limiting level. Thus, as against the case of the progressive wave where all the points along the axis Z aligned with a heat transfer surface are equally subjected to oscillations, in the considered case their essential nonuniformity (cellular periodicity) takes place. The noticed circumstance also results in a smaller degree of the thermal interaction "body – fluid" that in a quantitative sense results in an increase in the FC (i.e., to its smaller deviation from unity).

To summarize, it is possible to assert that the approximate algorithm of a computation of the FC constructed in the present section brings the research method on an essentially higher level. It removes the restrictions connected with individual consecutive computation of variants of the construction of the general problem solution with the purpose of obtaining the analytical solutions (harmonic, inverse harmonic, and symmetric step laws) and opens ample opportunities to investigate any periodic functions $\psi(\xi)$.

4.7 Filtration Property of the Computational Algorithm

The procedure of the exact solution of the heat conduction equation described in Chap. 3 included a calculation of a temperature field in an entire solid body, as well as a determination of the FC. This procedure uses a method of orthogonalization of the Fourier series and has a differential character. Therefore, such an approach inevitably results in extremely complex analytical solutions for the FC. On the other hand, heat transfer coefficients (both true, and experimental) determined as a result of the procedure of averaging by their very definition assume presence of some smoothed (integral or cumulative) properties. This intuitive assumption confirms the algorithm of calculation of the FC obtained above, which looks as a quadrature. As we deem, the integral form of the computational algorithm (4.20) should have as a consequence, in particular, a property of filtration of the high-frequency oscillations imposed on the basic functions $\psi(\xi)$. To be convinced in a validity of this assumption, we have investigated harmonic oscillations of the THTC modulated by a high-frequency component

$$\psi = \sin(\xi + n\sin(\xi)). \tag{4.74}$$

At increasing parameter n (i.e., at an increase in the frequency of the imposed oscillations), dependence $\varepsilon(\chi)$ approaches more and more closely to the corresponding dependence for the purely harmonic law of oscillations. The given particular example confirms basically the assumption of the filtration property of the computational algorithm. Unfortunately, we have not managed to find the strict proof of this extremely interesting and important assumption.

4.8 Generalized Parameter of the Thermal Effect

It is interesting to note that the approximate solutions (4.2) and (4.5) for a spatial problem coincide identically with a calculation based on the algorithm (4.20) (solutions (4.21) and (4.22)). One can be easily convinced in it assuming $m = 0$ in the latter equations. This fact is rather encouraging in the sense of the general ideology of direct variational methods [20]: a good agreement between the two independently obtained approximate solutions within the framework of the same method testifies in favor of the sufficient accuracy of each of these solutions. One should point out at the same time that the corresponding approximate solutions for the time-dependent problem obtained by two different methods nevertheless differ from each other, though this difference is insignificant. Obtained above with the help of a simplified solution of the heat conduction equation (2.7) was the relation (4.17) for the PTE. A comparative analysis of the approximate solutions obtained in the present Chapter allows writing down, instead of relation (4.17), the following expression

$$H = \sqrt{F_1^2 + \Phi_1^2}. \tag{4.75}$$

As shown in Appendix B, the dependence of functions F_1, Φ_1 on the wall thickness is characterized by decreasing oscillations imposed on the basic background. It means that the expression for the FC will also contain these oscillations. It is interesting to note that the only simple exact solution (3.84) obtained in Chap. 3 also contains an oscillating component decreasing at thickening of the wall.

4.9 Advantages of The Computational Algorithm

So, in our disposal there is an effective correct method for a calculation of the FC at any spatiotemporal oscillations of the THTC. Thus the applied objective of the theory of conjugate periodic heat transfer is achieved. An analysis of the problem can be presented schematically as the following logic chain:

- Initial conjugate convective–conductive problem "fluid flow – body" is replaced with a boundary problem for the heat conduction equation in a body with a BC of the third kind.
- As the BC, a THTC is preset that varies periodically around its average value with time and a coordinate along the heat transfer surface.
- A solution of the heat conduction equation for a considered general case of spatiotemporal oscillations [(2.3)] is represented as a superposition of the stationary [(2.4)] and fluctuation [(2.5)] components.
- A solution of the stationary equation is trivial [(2.6)]. The equation for temperature oscillations for the general case of oscillations of the THTC under the law of a progressive wave should be rewritten in the form of (2.7).
- Solutions of (2.7) for oscillations satisfying corresponding TBC [(2.8)] are written as (2.9).
- Unknown complex conjugate eigenvalues of the boundary problem, as well as the FC should be determined from the construction of the general solution [(3.7)–(3.9)].
- Analytical solutions for three characteristic functions of the THTC, harmonic, inverse harmonic, and symmetric stepwise, can be obtained. These solutions are very cumbersome (infinite complex conjugate chain fractions, infinite functional series). Analytical solutions for an arbitrary law of oscillations of the THTC do not exist.
- One should assume the simplified linear correlation (4.15) between oscillations of the temperatures and heat fluxes on a heat transfer surface. In view of this, from the BC in the form of (2.29) or (2.30) one can obtain the algorithm (4.20) for an approximate calculation of the FC

References

1. Zudin YB (1996) Theory on heat-transfer processes of periodic intensity. Habilitation. Moscow Power Engineering Institute (in Russian).
2. Zudin YB (1998) Temperature waves on a wall surface. Russ Dokl Phys J Acad Sci 43(5): 313–314.
3. Zudin YB (1999) Harmonic law of fluctuations of the true heat transfer coefficient. Thermophys Aeromech 6: 79–88.
4. Khinchin AY (1997) Continued Fractions. Dover Publications, New York.
5. Zudin YB (1996) Pulse law of true heat transfer coefficient pulsations. Appl Energy Russ J Fuel Power Heat Syst 34: 142–147.
6. Mikhailov MD, Özisik MN (1984) Unified Analysis and Solutions of Heat and Mass Diffusion. Wiley, New York.

7. Sommerfeld A (1978) Vorlesungen über Theoretische Physik, Band VI, Partielle Differentialgleichungen der Physik, Verlag Harry Deutsch, Frankfurt.
8. Zudin YB (1999) Some properties of the solution of the heat-conduction equation with periodic boundary condition of third kind. Thermophys Aeromech 6: 391–398.
9. Zudin YB (1998) Effect of the thermophysical properties of the wall on the heat transfer coefficient. Therm Eng 45(3): 206–209.
10. Sauer R, Szabo I (1969) Mathematische Hilfsmittel des Ingenieurs, Teil 1–4. Springer, Berlin Heidelberg New York.
11. Labuntsov DA (2000) Physical Principles of Energetics. Selected Papers. Power Engineering Institute, Moscow (in Russian).
12. Stephan K (1992) Heat Transfer in Condensation and Boiling. Springer, Berlin Heidelberg New York.
13. Kapitsa PL (1948) Wave flow of thin layers of a viscous liquid. Part 1. Free flow. Zh Eksp Teor Fiz 18(1): 1–28 (in Russian).
14. Kapitsa PL, Kapitsa SP (1949) Wave flow of thin layers of a viscous liquid. Part II. Fluid flow in the presence of continuous gas flow and heat transfer. Zh Eksp Teor Fiz 19(2): 105–120 (in Russian).
15. Carslaw HS, Jaeger JC (1992) Conduction of Heat in Solids. Clarendon Press, London, Oxford.
16. Stein EM, Shakarchi R (2003) Fourier Analysis: An Introduction. Princeton University Press, Princeton.
17. Abramovitz M, Stegun IA (1974) Handbook of Mathematical Functions with Formulas, Graphs, and Mathematical Tables. Dover Publications, New York.
18. Schlichting H, Gersten K (1997) Grenzschicht-Theorie. Springer, Berlin Heidelberg New York.
19. Hydrodynamic Instabilities and the Transition to Turbulence (1981) Swinney HL, Gollub JP (ed.). Springer, Berlin Heidelberg New York.
20. Kantorovich LV, Krylov VI (1958) Approximate Methods of Higher Analysis. P. Noordho, Groningen, The Netherlands.

5

Solution of Special Problems

As shown in Sect. 3.6, the dependence of the factor of conjugation (FC) on a wall thickness δ starts to exhibit itself explicitly when the value of δ becomes commensurate with the lengthscale of periodicity: $\delta \leq Z_0, \bar{\delta} \leq 1$. In a limiting case of the solely temporal fluctuations of the THTC $\left(m = Z_0^2/(\alpha \tau_0) \to \infty\right)$, a role of the lengthscale is played by the penetration length of the thermal wave $Z_0 \Rightarrow \sqrt{\alpha \tau_0}$, with the area of the wall thickness influence being determined with an inequality $\delta \leq \sqrt{\alpha t_0}, \tilde{\delta} \leq 1$. The dependence of the value of $\varepsilon(\delta)$ for two alternative thermal boundary conditions (TBC) has a mutually opposite character: a reduction of the wall thickness results in an increase in the FC at $\vartheta_0 = $ const and in a decrease in the FC at $q_0 = $ const (see Figs. 3.11 and 3.12). Considered in the present chapter are dependences of the function $\varepsilon(\delta)$ for the cases of more complex TBC such as:

- Stationary heat transfer at $X = 0$, TBC $h_0 = $ const
- Thermal contact to another (second) external solid body at $X = 0$, a condition of thermal conjugation

In this chapter, we also carry out a generalization of the results obtained above (for a flat plate) for bodies of other geometry, i.e., a cylinder and a sphere, with internal heat sources. Having at our disposal the universal computational algorithm developed in Chap. 4, it is possible to proceed to the solution of these special problems of the periodic conjugate heat transfer.

5.1 Complex Case of Heating or Cooling

Linear interrelation of fluctuations. In order to determine the PTE $- \chi = H/\langle \bar{h} \rangle$ [(4.15)] – let us use the simplified formula (4.17) that is very representative at exhibiting the results, though at the expense of some losses in accuracy. The corrected expressions corresponding to the case under consideration are documented in Appendix E. Let us present the expressions for fluctuations of

5 Solution of Special Problems

temperatures and heat fluxes for a complex case of the heat supply written in the following form:

$$\tilde{\vartheta} \approx \cos(\xi) \left(A_\vartheta \frac{\sinh(gx)}{\sinh(g\bar{\delta})} + A_q \frac{\cosh(gx)}{\cosh(g\bar{\delta})} \right), \tag{5.1}$$

$$\tilde{\vartheta}^\bullet \approx g \cos(\xi) \left(A_\vartheta \frac{\cosh(gx)}{\sinh(g\bar{\delta})} + A_q \frac{\sinh(gx)}{\cosh(g\bar{\delta})} \right), \tag{5.2}$$

where $g = (1 + m^2)^{1/4}$. Let us find from relations (5.1) and (5.2) the values of $\tilde{\vartheta}, \tilde{\vartheta}^\bullet$ at $X = 0$

$$\tilde{\vartheta} \approx \frac{A_q \cos(\xi)}{\cosh(g\bar{\delta})}, \quad \tilde{\vartheta}^\bullet \approx \frac{A_\vartheta g \cos(\xi)}{\sinh(g\bar{\delta})}. \tag{5.3}$$

Heat supply from an ambience. The condition of a stationary heat supply ($h_0 = \text{const}$) from an ambience (Fig. 5.1) with the temperature $\vartheta_\infty = \text{const}$ looks like

$$h_0 (\vartheta_\infty - \vartheta_0) = -k \left. \frac{\partial \vartheta}{\partial X} \right|_{X=0}. \tag{5.4}$$

Owing to linearity of the condition (5.4), it will also hold for the fluctuation component of the temperature

$$\bar{h}_0 \hat{\vartheta}_0 = \hat{\vartheta}_0^\bullet, \tag{5.5}$$

where $\bar{h}_0 = h_0 Z_0/k$. It can be obtained from relations (5.3) and (5.5) that

$$\frac{A_\vartheta}{A_q} = \frac{\bar{h}_0 \tanh(g\bar{\delta})}{g}. \tag{5.6}$$

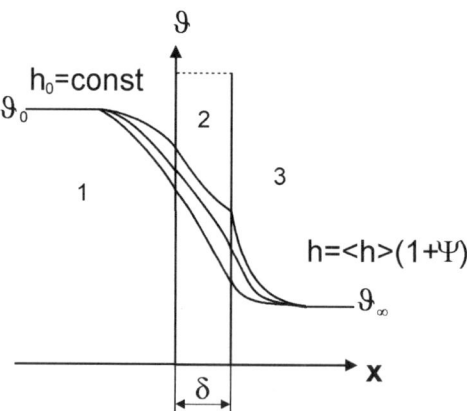

Fig. 5.1. Heat supply from the ambience: (**1**) ambience, (**2**) body and (**3**) cooling fluid

5.1 Complex Case of Heating or Cooling

Assuming $X = \delta$ in (5.1) and (5.2), one can find, with account for (5.6), the value $H = \tilde{\vartheta}_\delta^\bullet / \tilde{\vartheta}_\delta$ and further the PTE from the formula (4.17)

$$\chi = \frac{H}{\langle \bar{h} \rangle} = \frac{g}{\langle \bar{h} \rangle} \frac{\bar{h}_0 + g \tanh(g\bar{\delta})}{\bar{h}_0 \tanh(g\bar{\delta}) + g}. \tag{5.7}$$

Using the relation (5.7) in corresponding solutions for the FC, one can obtain an approximate solution of the problem of periodic conjugate heat transfer with a stationary external heat supply. Tendencies of the dependence of the FC on determining parameters for the considered multiparameter case are listed below:

- The general tendencies of the wall's thermal effect expressed by the Biot number remain in force

$$\left. \begin{array}{l} \langle \bar{h} \rangle \to 0 : \varepsilon \to 1, \\ \langle \bar{h} \rangle \to \infty : \varepsilon \to \varepsilon_{\min}. \end{array} \right\} \tag{5.8}$$

Asymptotical solutions (5.8) reflect limiting cases $k \to \infty$ and $k \to 0$, respectively. They are natural, since the character of the thermal effect of a basic body on the heat transfer characteristics does not change with a complication of the TBC.

- The character of the Fourier number's influence, which determines the interaction of the spatial and temporal periodicity of fluctuations, also holds

$$\left. \begin{array}{l} m \to 0 : \chi = \frac{1}{\langle \bar{h} \rangle} \frac{\bar{h}_0 + \tanh(\bar{\delta})}{\bar{h}_0 \tanh(\bar{\delta}) + 1} \\ m \to \infty : \chi = \frac{1}{\langle \bar{h} \rangle} \frac{\bar{h}_0 + \tanh(\bar{\delta})}{\bar{h}_0 \tanh(\bar{\delta}) + 1} \end{array} \right\} \tag{5.9}$$

It is apparent from (5.9) that dependences $\varepsilon(\delta)$ in both limiting cases $Z_0 \to 0$ also $\tau_0 \to 0$ are identical. Limiting TBC correspond to the limiting values of the external stationary heat transfer coefficient (HTC)

$$\left. \begin{array}{l} \bar{h}_0 \to \infty : \chi = \frac{g}{\langle \bar{h} \rangle} \coth(g\bar{\delta}) \Rightarrow \vartheta_0 = \text{const}, \\ \bar{h}_0 \to 0 : \chi = \frac{g}{\langle \bar{h} \rangle} \tanh(g\bar{\delta}) \Rightarrow q_0 = \text{const} \end{array} \right\} \tag{5.10}$$

Thus, the complex TBC $h_0 = \text{const}$ represents a natural generalization of the two simple TBC considered earlier.

- At transition to a semi-infinite body, the influence of the TBC degenerates

$$\bar{\delta} \to \infty : \chi = \frac{g}{\langle \bar{h} \rangle}. \tag{5.11}$$

This asymptotical solution reflects a natural tendency of forgetting the TBC at transition from a finite plate to a semi-infinite body.

- A limiting case of a plate with a vanishingly small thickness is described by the asymptotical solution

$$\bar{\delta} \to 0 : \chi = \frac{h_0}{\langle h \rangle}. \tag{5.12}$$

For the case of harmonic fluctuations determined by the relation (3.10), the FC is given by the following relation:

$$\varepsilon \approx \sqrt{\varepsilon_{\min}^2 + 2\frac{h_0}{\langle h \rangle} + \left(\frac{h_0}{\langle h \rangle}\right)^2} - \frac{h_0}{\langle h \rangle}. \tag{5.13}$$

A remarkable feature of the relation (5.13) is that the dimensionless determining parameters, namely, Biot and Fourier numbers, are absent in it. Indeed, if the solid wall disappears, its thermophysical properties (thermal conductivity k and thermal diffusivity α), naturally, cannot influence on the thermal interaction of two fluids flowing over the two sides of the wall: external (stationary) and internal (disturbed). The wall thickness effect in this case also degenerates. Let us find out how the influence of the external HTC exhibits itself in this case. At an infinitely large external heat supply intensity $h_0/\langle h \rangle \to \infty$, the relation (5.13) describes the limiting nonconjugate case $\varepsilon = 1$. On the contrary, a negligibly small level of stationary heat transfer ($h_0/\langle h \rangle \to 0$) corresponds to the case of the maximal effect of the conjugation $\varepsilon = \varepsilon_{\min}$. Thus, even at absence of a wall, the thermal conjugation (this time between two fluids) nonetheless takes place.

At last, let us point out at the known analogy to a classical problem of thermal conductivity describing a nonstationary field of temperatures in a semi-infinite body with a boundary condition (BC) of the third kind [1]. Here also limiting TBC correspond to the limiting values of a constant HTC on a surface of a body (see Sect. 1.4).

– An interesting case is realized under the condition of $\bar{h}_0 = g$, a developed notation of which looks like

$$\left(\frac{h_0 Z_0}{k}\right)^4 = 1 + \left(\frac{Z_0^2}{\alpha \tau_0}\right)^2. \tag{5.14}$$

As follows from the relations (5.11) and (5.12), for this "equilibrium case" the PTE does not depend on the wall thickness and it is equal to its correspondent value at $\delta \to \infty$ (Fig. 5.2). This result most evidently exhibits itself at a transition from the general case of spatiotemporal fluctuations of the THTC to their limiting (with respect to the Fourier number) forms

$$\left.\begin{array}{l} m \to 0 : \frac{h_0 Z_0}{k} = 1, \quad h_0 = \frac{k}{Z_0}, \\ m \to \infty : \frac{h_0 \sqrt{\alpha \tau_0}}{k} = 1, \quad h_0 = \sqrt{\frac{kc\rho}{\tau_0}}. \end{array}\right\} \tag{5.15}$$

Thus, the "equilibrium" values of the stationary HTC are determined by spatial and temporal scales of fluctuations and also by thermophysical properties of a solid body.

Thermal contact to another (second) body. A case of the thermal contact to another (second) body (Fig. 5.3) is described by stationary conditions of

5.1 Complex Case of Heating or Cooling

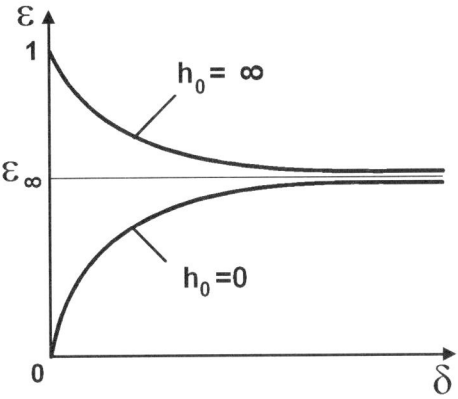

Fig. 5.2. Equilibrium case of the external heat transfer

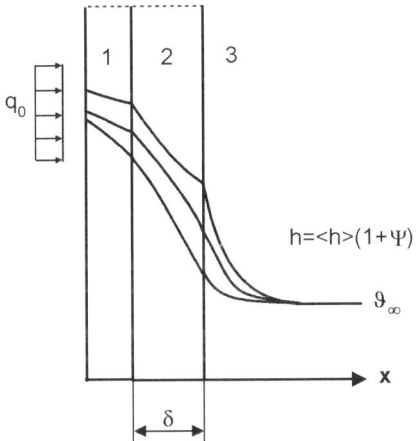

Fig. 5.3. Heat supply from an external body: (1) external body, (2) body, (3) ambient fluid

the conjugation: equality of temperatures and heat fluxes on the interface of the contact [2]

$$\vartheta_0 = \vartheta_w, \quad q_0 = q_w. \tag{5.16}$$

Owing to the linearity of (5.16), they hold also for the fluctuation component of the temperature

$$\tilde{\vartheta}_0 = \tilde{\vartheta}_w, \quad \tilde{\vartheta}_0^\bullet = \frac{g_w}{g} \frac{k_w}{k} \tilde{\vartheta}_w^\bullet. \tag{5.17}$$

Here $g = (1 + m^2)^{1/4}$, $g_w = [1 + m^2 (\alpha/\alpha_w)]^{1/4}$, the subscript "$w$" denotes conditions of the second (external) wall. One can further obtain from relations (5.3) and (5.17)

$$\frac{A_\vartheta}{A_q} = F_w \frac{g_w}{g} \frac{k_w}{k} \tanh\left(g\bar{\delta}\right). \tag{5.18}$$

Here F_w is a function of the thickness of the second wall, corresponding to the standard TBC

$$\left. \begin{array}{l} \vartheta_{w0} = \text{const} : F_w = \tanh\left(g_w \bar{\delta}_w\right), \\ q_{w0} = \text{const} : F_w = \coth\left(g_w \bar{\delta}_w\right). \end{array} \right\} \quad (5.19)$$

The value of the PTE can be determined from (5.1) and (5.2)

$$\chi = \frac{g}{\langle \bar{h} \rangle} \frac{(g_w k_w)/(gk) F_w + \tanh\left(g\bar{\delta}\right)}{(g_w k_w)/(gk) F_w \tanh\left(g\bar{\delta}\right) + 1}. \quad (5.20)$$

The considered problem of periodic heat transfer for a two-layer wall incorporates a plenty of the individual problems realized in asymptotical solutions for the individual parameters of this problem.

- The influence of the Biot number is still described by (5.8)
- An asymptotical solution for the Fourier number takes the following form:

$$\left. \begin{array}{l} m \to 0 : \chi = \dfrac{1}{\langle \bar{h} \rangle} \dfrac{(k_w/k) F_w + \tanh(\bar{\delta})}{(k_w/k) F_w \tanh(\bar{\delta}) + 1}, \\[2mm] m \to \infty : \chi = \dfrac{1}{\langle \bar{h} \rangle} \dfrac{\sqrt{(k_w c_w \rho_w)/(kc\rho)} F_w + \tanh(\bar{\delta})}{\sqrt{(k_w c_w \rho_w)/(kc\rho)} F_w \tanh(\bar{\delta}) + 1}. \end{array} \right\} \quad (5.21)$$

One should point out that for the limiting case of the temporal fluctuations $m \to \infty$, the function of the thickness of the second wall looks like

$$\left. \begin{array}{l} \vartheta_{w0} = \text{const} : F_w = \tanh\left(\tilde{\delta}_w\right), \\ q_{w0} = \text{const} : F_w = \coth\left(\tilde{\delta}_w\right), \end{array} \right\} \quad (5.22)$$

where $\tilde{\delta}_w = \delta_w / \sqrt{\alpha_w \tau_0}$.

- Standard TBC (5.10) correspond to limiting values of thermal conductivity of the external wall

$$\left. \begin{array}{l} \dfrac{k_w}{k} \to \infty : \chi = \dfrac{g}{\langle \bar{h} \rangle} \coth\left(g\bar{\delta}\right) \Rightarrow \vartheta_0 = \text{const}, \\[2mm] \dfrac{k_w}{k} \to 0 : \chi = \dfrac{g}{\langle \bar{h} \rangle} \tanh\left(g\bar{\delta}\right) \Rightarrow q_0 = \text{const} \end{array} \right\} \quad (5.23)$$

- For the case of temporal fluctuations $m \to \infty$, an obvious redefinition of the parameter of influence should be done $k_w/k \Rightarrow \sqrt{(k_w c_w \rho_w)/(kc\rho)}$, and the system of (5.23) takes a resulting form

$$\left. \begin{array}{l} \sqrt{\dfrac{k_w c_w \rho_w}{kc\rho}} \to \infty : \chi = \dfrac{1}{\langle \tilde{h} \rangle} \coth\left(\tilde{\delta}\right) \Rightarrow \vartheta_0 = \text{const}, \\[2mm] \sqrt{\dfrac{k_w c_w \rho_w}{kc\rho}} \to 0 : \chi = \dfrac{1}{\langle \tilde{h} \rangle} \tanh\left(\tilde{\delta}\right) \Rightarrow q_0 = \text{const} \end{array} \right\} \quad (5.24)$$

Here $\langle \tilde{h} \rangle = \langle h \rangle \sqrt{\alpha \tau_0}/k$, $\tilde{\delta} = \delta/\sqrt{\alpha \tau_0}$.

5.1 Complex Case of Heating or Cooling

– At an infinite increase in the thickness of the second body $\delta_w \to \infty$, we have $F_w \to 1$. In this case, dependence $\chi(\delta_w)$ degenerates, and the PTE can be written as

$$\chi = \frac{g}{\langle \bar{h} \rangle} \frac{(g_w k_w)/(gk) + \tanh(g\bar{\delta})}{(g_w k_w)/(gk) \tanh(g\bar{\delta}) + 1}. \tag{5.25}$$

It is interesting to note that the examined case of an external contact with a semi-infinite body is similar to the case of external heat transfer considered above. Indeed, (5.25) can be easily derived from (5.7) at the following replacement $\bar{h}_0 \Rightarrow g_w (k_w/k)$. Based on this replacement, one can deduce the relations equivalent to (5.10), (5.23), and (5.24)

$$\left. \begin{array}{l} \frac{g_w}{g} \frac{k_w}{k} \to \infty : \chi = \frac{g}{\langle \bar{h} \rangle} \coth(g\bar{\delta}) \Rightarrow \vartheta_0 = \mathrm{const}, \\ \frac{g_w}{g} \frac{k_w}{k} \to 0 : \chi = \frac{g}{\langle \bar{h} \rangle} \tanh(g\bar{\delta}) \Rightarrow q_0 = \mathrm{const} \end{array} \right\} \tag{5.26}$$

The intermediate "equilibrium" case with $\chi \neq \chi(\delta)$, where the FC does not depend on the plate thickness can be realized here under the following condition:

$$\frac{g_w k_w}{gk} = 1, \tag{5.27}$$

whose extended form of notation looks like

$$\frac{k_w}{k} = \left[\frac{1 + (Z_0/\sqrt{\alpha \tau_0})^4}{1 + (Z_0/\sqrt{\alpha_w \tau_0})^4} \right]^{1/4}. \tag{5.28}$$

In the limiting cases (with respect to the Fourier number), the "equilibrium" conditions are described by the physically obvious relations

$$\left. \begin{array}{l} m \to 0 : \frac{k_w}{k} = 1, \\ m \to \infty : \sqrt{\frac{k_w c_w \rho_w}{kc\rho}} = 1. \end{array} \right\} \tag{5.29}$$

– Let us consider now an asymptotical case alternative to the previously examined one. At $\bar{\delta}_w \to 0$, the entire number of parameters of an external body can be reduced to the task of specifying the TBC on a contact surface. In other words, the standard TBC are transferred from the external surface of the second wall with a vanishingly small thickness onto the external surface of the basic wall. Thus, we return back to the considered above cases of the simple TBC. This asymptotical case is similar to the corresponding asymptotical solution (5.14) for the case of a stationary external HTC.
– One more pair of asymptotical solutions is realized at $\delta \to 0$. It is physically obvious, that an internal body of a vanishingly small thickness will not affect the characteristics of heat transfer. We again return here to the standard TBC, however realized already for the second (external) body.

Thus, a transition from the standard TBC to a complex case of an external heat supply considerably enriches a spectrum of individual subproblems of a general conjugate heat transfer problem and once again emphasizes the advantages of the approximate analytical solutions. It is also important to note that the case of a thermal contact to the second body has one additional parameter in comparison with the case of a heat supply from an ambience. It can be explained physically by such a reasoning that the influence of an external body on heat transfer characteristics exhibits itself not only through the thermophysical properties of the body, but also through its thickness. For a case where the second body is semi-infinite, its thickness, naturally, ceases to affect the heat transfer characteristics. Thus both variants of the complex heat supply become equal in the number of parameters.

A rather interesting conclusion follows from an analysis of the limiting case of the zero thickness of an internal body. In this case, the conjugation effect ceases to depend on the internal body's properties, however, it holds anyway, and the role of a damping wall is taken over by a flow of an external medium (or by an external body).

5.2 Heat Transfer on the Surface of a Cylinder

The case of periodic heat transfer on a surface of the cylinder with internal heat sources is considered similarly to the purely time-dependent problem investigated in Sect. 2.1 for a flat plate with the TBC $q_0 = \text{const}$. The heat conduction equation for temperature fluctuations takes here the following form [1]:

$$\frac{\partial \tilde{\vartheta}}{\partial t} = \frac{1}{\tilde{r}} \frac{\partial}{\partial \tilde{r}} \left(\tilde{r} \frac{\partial \tilde{\vartheta}}{\partial \tilde{r}} \right), \tag{5.30}$$

where $t = \tau/\tau_0, \tilde{r} = r/\sqrt{a\tau_0}$, r is the radial coordinate counted from the axis of symmetry of the cylinder. An application of the method of separation of variables [3] to (5.30) results in the following solution:

$$\tilde{\vartheta} = \sum_{n=1}^{\infty} \left[A_n \frac{\mathrm{ber}_0\left(\sqrt{n}\tilde{r}\right) + i\,\mathrm{bei}_0\left(\sqrt{n}\tilde{r}\right)}{\mathrm{ber}_0\left(\sqrt{n}\tilde{R}\right) + i\,\mathrm{bei}_0\left(\sqrt{n}\tilde{R}\right)} \exp(int) \right.$$
$$\left. + A_n^* \frac{\mathrm{ber}_0\left(\sqrt{n}\tilde{r}\right) - i\,\mathrm{bei}_0\left(\sqrt{n}\tilde{r}\right)}{\mathrm{ber}_0\left(\sqrt{n}\tilde{R}\right) - i\,\mathrm{bei}_0\left(\sqrt{n}\tilde{R}\right)} \exp(-int) \right]. \tag{5.31}$$

Here $\tilde{R} = R/\sqrt{a\tau_0}$, R is the outer radius of the cylinder, $\mathrm{ber}_0(x), \mathrm{bei}_0(x)$, $\mathrm{ber}_1(x), \mathrm{bei}_1(x)$, are Thomson's functions [4]. In this case, expressions for fluctuation components of temperatures and temperature gradients at $r = R$ are given as

$$\tilde{\vartheta}_R = \sum_{n=1}^{\infty} [A_n \exp(\mathrm{i} n t) + A_n^* \exp(-\mathrm{i} n t)], \qquad (5.32)$$

$$\tilde{\vartheta}_R^{\bullet} = \sum_{n=1}^{\infty} [B_n A_n \exp(\mathrm{i} n t) + B_n^* A_n^* \exp(-\mathrm{i} n t)]. \qquad (5.33)$$

Here $B_n = F_n + \mathrm{i}\Phi_n$, $B_n^* = F_n - \mathrm{i}\Phi_n$ are complex conjugate eigenvalues of the boundary problem, and

$$\left.\begin{aligned}
F_n &= \sqrt{\tfrac{n}{2}}\, \tfrac{\mathrm{ber}_0(\Delta)[\mathrm{bei}_1(\Delta)-\mathrm{ber}_1(\Delta)]+\mathrm{ber}_0(\Delta)[\mathrm{bei}_1(\Delta)+\mathrm{ber}_1(\Delta)]}{\mathrm{bei}_0^2(\Delta)+\mathrm{ber}_0^2(\Delta)}, \\
\Phi_n &= \sqrt{\tfrac{n}{2}}\, \tfrac{\mathrm{ber}_0(\Delta)[\mathrm{bei}_1(\Delta)+\mathrm{ber}_1(\Delta)]+\mathrm{ber}_0(\Delta)[\mathrm{bei}_1(\Delta)-\mathrm{ber}_1(\Delta)]}{\mathrm{bei}_0^2(\Delta)+\mathrm{ber}_0^2(\Delta)}, \\
\Delta &= \sqrt{n}\tilde{R}.
\end{aligned}\right\} \qquad (5.34)$$

Thus, the field of temperature fluctuations on the surface of a cylinder is equivalent to the correspondent problem of temporal fluctuations of the heat transfer intensity on a flat plate surface, with the functions of thickness F_n, Φ_n. being redefined using (5.34).

5.3 Heat Transfer on the Surface of a Sphere

The heat conduction equation for temperature fluctuations in a sphere has the following form [1]:

$$\frac{\partial \tilde{\vartheta}}{\partial t} = \frac{1}{\tilde{r}^2}\frac{\partial}{\partial \tilde{r}}\left(\tilde{r}^2 \frac{\partial \tilde{\vartheta}}{\partial \tilde{r}}\right). \qquad (5.35)$$

A solution of (5.35) looks as

$$\tilde{\vartheta} = \frac{R}{r}\sum_{n=1}^{\infty}\left\{ A_n \frac{\mathrm{sh}\left[\sqrt{n/2}\,(1+\mathrm{i})\,\tilde{r}\right]}{\mathrm{sh}\left[\sqrt{n/2}\,(1+\mathrm{i})\,\tilde{R}\right]} \exp(\mathrm{i} n t) \right.$$

$$\left. + A_n^* \frac{\mathrm{sh}\left[\sqrt{n/2}\,(1-\mathrm{i})\,\tilde{r}\right]}{\mathrm{sh}\left[\sqrt{n/2}\,(1-\mathrm{i})\,\tilde{R}\right]} \exp(-\mathrm{i} n t) \right\}. \qquad (5.36)$$

An expression for the temperature fluctuations on the surface of a sphere (at $r = R$) can be written down similarly to (5.32)

$$\tilde{\vartheta}_R = \sum_{n=1}^{\infty} [A_n \exp(\mathrm{i} n t) + A_n^* \exp(-\mathrm{i} n t)]. \qquad (5.37)$$

Fluctuations of heat fluxes at $r = R$ can be presented as

$$\tilde{\vartheta}_R^{\bullet} = \tilde{\vartheta}_\delta^{\bullet} - \frac{\tilde{\vartheta}_R}{R}, \qquad (5.38)$$

where $\tilde{\vartheta}_\delta^{\bullet}$ is the value characteristic for the case of a plate with the TBC $\vartheta_0 = \mathrm{const}$.

5.4 Parameter of Thermal Effect for Different Geometrical Bodies

Functions of the thickness for a plate. Let us determine, from relations (3.44) and (3.45), an interrelation of root-mean-square fluctuations of the temperatures and temperature gradients at $X = \delta$

$$H^2 \equiv \frac{\langle\theta^{\bullet 2}\rangle}{\langle\theta^2\rangle} = \frac{\sum_{n=1}^{\infty}\left[\left(F_n^2 + \Phi_n^2\right)\left(R_n^2 + I_n^2\right)\right]}{\sum_{n=1}^{\infty}\left(R_n^2 + I_n^2\right)}. \tag{5.39}$$

Based on the properties of functions $B_n = F_n + i\Phi_n$, $B_n^* = F_n - i\Phi_n$ (see Appendix B), one can prove the validity of an inequality

$$F_1^2 + \Phi_1^2 \leq F_n^2 + \Phi_n^2, \quad n = 1, 2, 3, \ldots. \tag{5.40}$$

From (5.39) and (5.40) follows also validity of an inequality

$$\langle\theta^{\bullet 2}\rangle \geq H_1^2 \langle\theta^2\rangle. \tag{5.41}$$

Function H_1 can be determined by a relation

$$H_1 = \sqrt{F_1^2 + \Phi_1^2}, \tag{5.42}$$

where

$$\left.\begin{array}{l} F_1 = \frac{1}{\sqrt{2}} \frac{\sinh(\sqrt{2}\tilde{\delta}) - \sin(\sqrt{2}\tilde{\delta})}{\cosh(\sqrt{2}\tilde{\delta}) + \cos(\sqrt{2}\tilde{\delta})}, \\ \Phi_1 = \frac{1}{\sqrt{2}} \frac{\sinh(\sqrt{2}\tilde{\delta}) + \sin(\sqrt{2}\tilde{\delta})}{\cosh(\sqrt{2}\tilde{\delta}) + \cos(\sqrt{2}\tilde{\delta})}. \end{array}\right\} \tag{5.43}$$

Let us replace the inequality (5.41) with a chain of the following approximate relations:

$$|\theta^{\bullet 2}| \geq H_1^2 |\theta^2|, \quad \theta^{\bullet 2} \geq H_1 \theta^2, \quad \theta^{\bullet} \approx H_1 \theta. \tag{5.44}$$

For the sake of convenience, let us agree to write below throughout H instead of H_1. Expressions for the function H for a flat plate are determined by the relation (4.17) from Sect. 4.3. Correspondent relations for the cases of a cylinder and a sphere are given below.

Functions of the thickness for a cylinder. Function H_1 is determined by the very same (5.42). Functions F_1, Φ_1 can be obtained assuming $n = 1$ in (5.34)

$$\left.\begin{array}{l} F_1 = \frac{1}{\sqrt{2}} \frac{\mathrm{ber}_0(\tilde{R})\left[\mathrm{bei}_1(\tilde{R}) - \mathrm{ber}_1(\sqrt{\tilde{R}})\right] + \mathrm{bero}(\tilde{R})\left[\mathrm{bei}_1(\tilde{R}) + \mathrm{ber}_1(\tilde{R})\right]}{\mathrm{bei}_0^2(\tilde{R}) + \mathrm{ber}_0^2(\tilde{R})}, \\ \Phi_1 = \frac{1}{\sqrt{2}} \frac{\mathrm{ber}_0(\tilde{R})\left[\mathrm{bei}_1(\tilde{R}) + \mathrm{ber}_1(\tilde{R})\right] + \mathrm{bero}(\tilde{R})\left[\mathrm{bei}_1(\tilde{R}) - \mathrm{ber}_1(\tilde{R})\right]}{\mathrm{bei}_0^2(\tilde{R}) + \mathrm{ber}_0^2(\tilde{R})}. \end{array}\right\} \tag{5.45}$$

Functions of the thickness for a sphere. With an account for (5.38), function H_1 can be determined by the equation

$$H_1 = \sqrt{(F_1 - 1/R)^2 + \Phi_1^2}, \qquad (5.46)$$

where

$$\left. \begin{array}{l} F_1 = \dfrac{1}{\sqrt{2}} \dfrac{\sinh(\sqrt{2}\tilde{R}) + \sin(\sqrt{2}\tilde{R})}{\cosh(\sqrt{2}\tilde{R}) - \cos(\sqrt{2}\tilde{R})}, \\[6pt] \Phi_1 = \dfrac{1}{\sqrt{2}} \dfrac{\sinh(\sqrt{2}\tilde{R}) - \sin(\sqrt{2}\tilde{R})}{\cosh(\sqrt{2}\tilde{R}) - \cos(\sqrt{2}\tilde{R})}. \end{array} \right\} \qquad (5.47)$$

A generalized solution for a plate, a cylinder and a sphere. The generalized heat conduction equation for temperature fluctuations for bodies of a standard form (plate, cylinder, and sphere) can be written in the following form [1]:

$$\frac{\partial \tilde{\vartheta}}{\partial t} = \frac{1}{\tilde{x}^s} \frac{\partial}{\partial x} \left(\tilde{x}^s \frac{\partial \tilde{\vartheta}}{\partial x} \right). \qquad (5.48)$$

Here $t = \tau/\tau_0$, $\tilde{x} = X/\sqrt{a\tau_0}$; x is the cross-section coordinate counted from the plane of symmetry (for a plate); the axis of symmetry (for a cylinder), the center of symmetry (for a sphere); s is the geometrical factor equal to $s = 0$ for a plate, $s = 1$ for a cylinder, $s = 2$ for a sphere. According to (5.48), the general form of the expression for the PTE can be simplified as

$$\chi \approx \frac{1}{\langle \tilde{h} \rangle} \tanh\left(\frac{\tilde{x}}{1+s} \right). \qquad (5.49)$$

Using the formula in algorithm (4.20) for the preset type of fluctuations of the THTC, one can obtain a generalized approximate solution of the problem for bodies of the standard form. In this case, the cross-section coordinate is determined from the generalized relation s as a quotient of a division of a body's volume by the surface area of heat transfer.

5.5 Overall ATHTC

5.5.1 Overall EHTC

A three-part chain of the conjugate heat transfer. Everywhere above we considered a case of single-sided convective heat transfer. In that case, differences between the ATHTC and EHTC can be formally treated as a result of an application of different procedures of averaging of the HTC under conditions of its periodic fluctuations. With a reference to the engineering applications, it means the following. Let us, for example, assume that on the external surface of a body at $X = 0$ the TBC $q_0 = $ const is specified. Then a quantitative change of the

EHTC due to the thermal effect of a solid body will lead to a change of the average temperature difference $\langle \vartheta_\delta \rangle = \langle q \rangle / h_m$ on the internal surface of the body at $X = \delta$. This fact itself does not contain essentially new information. An essentially different situation will take place by consideration of two-sided convective heat transfer: at $X = 0$, a stationary TBC of the third kind is preset; it is required to find out an overall EHTC

$$\frac{1}{U_m} = \frac{1}{h_0} + \frac{\delta}{k} + \frac{1}{h_m}. \tag{5.50}$$

The parameter U_m determines an averaged (over the period of fluctuations) heat flux, which is transferred through the three-part system "stationary convective heat transfer – heat conduction – fluctuation convective heat transfer"

$$\langle q \rangle = U_m \vartheta_\Sigma. \tag{5.51}$$

Here ϑ_Σ is the total temperature difference in the three-part system. From an obvious condition of the average thermal balance over the period of fluctuations, the value of $\langle q \rangle$ for each of the thermal parts of the system should remain constant. It follows directly from here that a decrease in the value of h_m due to the wall's thermal effect will result, accordingly to (5.50), in a correspondent decrease in the value of U_m. Therefore, at a fixed full temperature difference ($\vartheta_\Sigma = \text{const}$), the value of $\langle q \rangle$ accordingly to (5.51) will also decrease.

An optimum wall thickness. Let us consider now the second important conclusion following from the thermal interaction in the fames of the problem "fluid – body." In a practice of the calculation of heat exchangers [5], it is traditionally deemed that a decrease in the wall thickness automatically leads to a decrease in the total thermal resistance of the three-part system. However, it is not always true in view of the thermal effect of a body. At a certain combination of the thermal resistances participating in the general chain, certain situations are principally possible at which a reduction of the wall thickness can lead to an increase in the total thermal resistance of the three-part systems.

Let us consider the symmetric step law of a variation of the THTC. With a purpose to simplify the calculations and to render a presentation of the results a more obvious from, we shall consider a problem of the solely spatial fluctuations of the heat transfer intensity along the internal surface. The fluctuation component of the THTC can be specified in this case by the formula

$$\left. \begin{array}{l} 0 \leq z \leq \frac{Z_0}{2} : \psi = 1, \\ \frac{Z_0}{2} \leq z \leq Z_0 : \psi = -1, \end{array} \right\} \tag{5.52}$$

where Z_0 is the lengthscale of the spatial periodicity (the length of the wave of fluctuations). Let us limit ourselves with a rather rough approximation of the effect of the dependence of EHTC on the wall thickness expressed by the TBC $h_0 = \text{const}$

$$\varepsilon = \frac{1}{1 + F \langle \bar{h} \rangle}. \tag{5.53}$$

5.5 Overall ATHTC

Here
$$F = \frac{1 + (1 + \bar{h}_0)\bar{\delta}}{\bar{h}_0 + (1 + \bar{h}_0)\bar{\delta}} - \qquad (5.54)$$

is a function of the wall thickness; $\langle \bar{h} \rangle = \langle h \rangle Z_0/k$, $\bar{h}_0 = h_0 Z_0/k$ are the Biot numbers for the internal and external surfaces, respectively; $\bar{\delta} = \delta/Z_0$ is the dimensionless wall thickness. A substitution of (5.53) and (5.54) into (5.51) with an allowance for (2.13) for the FC gives the following relation for the dimensionless overall EHTC:

$$\frac{1}{\bar{U}_m} = \frac{1}{\langle \bar{h} \rangle} + \frac{1}{\bar{h}_0} + F_1. \qquad (5.55)$$

Here $F_1 = F + \bar{\delta}$. The dimensionless overall THTC determined in absence of the thermal effect of a wall is then equal to

$$\frac{1}{\langle \bar{U} \rangle} = \frac{1}{\langle \bar{h} \rangle} + \frac{1}{\bar{h}_0} + \bar{\delta}. \qquad (5.56)$$

The function F_1 reaches its minimum at the value of $\bar{\delta}$ determined by the relation

$$\bar{\delta}_* = \frac{\sqrt{1 - \bar{h}_0^2} - \bar{h}_0}{1 + \bar{h}_0}. \qquad (5.57)$$

Relation (5.57) allows determining the range of a possible variation of parameters at the point of the minimum of the function F_1

$$0 < \bar{h}_0 < 1/\sqrt{2} : 0 < \bar{\delta}_* < 1. \qquad (5.58)$$

Using expressions (5.58) in (5.55) and (5.56), one can find out corresponding relations for the conditions of a maximum of the parameter $\bar{U}_m$

$$\frac{1}{\bar{U}_{m*}} = \frac{1}{\langle \bar{h} \rangle} + \frac{1}{\bar{h}_0} + \frac{1 + 2\sqrt{1 - \bar{h}_0^2}}{1 + \bar{h}_0}, \qquad (5.59)$$

$$\frac{1}{\langle \bar{U}_* \rangle} = \frac{1}{\langle \bar{h} \rangle} + \frac{1}{\bar{h}_0} + \frac{\sqrt{1 - \bar{h}_0^2} - \bar{h}_0}{1 + \bar{h}_0}. \qquad (5.60)$$

As follows from relations (5.59) and (5.60), the dependence $\bar{U}_{m*}(\bar{\delta})$ can reach its maximum in rather narrow ranges of the variation of the determining parameters. However, in this case the influence of the thermal effect of a body on the averaged heat transfer can be itself rather essential. It is interesting to introduce (by analogy to the case of single-sided heat transfer) a generalized factor of conjugation for the considered case of two-sided heat transfer

$$E = \frac{U_m}{\langle U \rangle}. \qquad (5.61)$$

108   5 Solution of Special Problems

An analysis of the problem of two-sided convective heat transfer under conditions of the thermal effect of a solid body brings the method presented in this book on an essentially new hierarchical level. In particular, essentially novel opportunities open for an optimization of heat exchangers with respect to the wall thickness.

Issues of the heat transfer intensification. One more interesting applied aspect of the three-part problem of the conjugate heat transfer is an issue of the heat transfer intensification due to the imposition of external fluctuations on a stationary heat transfer background. As shown in works [5–8], this results in an increase of the average heat transfer level. One of the possible realizations of an interaction of the stationary and fluctuation components of the THTC was considered in Chap. 1 at the analysis of the fluctuation laminar boundary layer (Sect. 1.3, (1.12)–(1.16)). According to the general concept of the thermal effect of a body stated in the present book, an increase in the amplitude of fluctuations of the THTC entails a decrease in the FC and, as a consequence, to a decrease in the value of U_m. As a result, the expected effect of the intensification will be compensated to some extent by the influence of the thermal effect of a wall. One should take this novel effect into account at a solution of particular problems of heat transfer intensification and, in particular, include into these solutions certain additional amendments allowing for the heat transfer deterioration.

5.5.2 Bilateral Spatiotemporal Periodicity of Heat Transfer (A Qualitative Analysis)

Let us consider now a case of bilateral spatiotemporal periodicity of heat transfer: on the left-hand side of a plate, spatial fluctuations are preset, while on the right-hand side, temporal fluctuations of the heat transfer intensity are imposed. If the wall thickness is much larger than both linear lengthscales $\left(\delta \gg Z_0, \delta \gg \sqrt{\alpha \tau_0}\right)$, then for each of the sides of heat transfer there will be its own dependence of the EHTC on the Biot number corresponding to the case of a semi-infinite body. If a wall thickness is commensurate even with just one linear lengthscale, a mutual influence of two mechanisms of fluctuations takes place. This case is much more complex in comparison with the considered above single-sided case of spatiotemporal periodicity such as the progressive wave, so that it is not obviously possible here to obtain a strict analytical solution of the problem. Since, however, the mentioned crisscross imposition of two types of periodicity is breathtakingly interesting, we believed necessary to carry out a qualitative analysis of this problem. Like we have done above, let us consider the symmetric step law of the variation of the THTC on both sides of a wall (see (5.52)). Let us accept roughly that a linear superposition of the corresponding "conjugate" thermal resistances takes place. Then, instead of (5.50), one can obtain

$$\frac{1}{U_\mathrm{m}} = \frac{\delta}{k} + \frac{1}{\langle h_z \rangle} + \frac{1}{\langle h_t \rangle} + R_z + R_t, \qquad (5.62)$$

5.5 Overall ATHTC

where $\langle h_z \rangle$, $\langle h_t \rangle$ are corresponding values of the ATHTC for each of the sides of heat transfer. An additional thermal resistance caused by the thermal conjugation can be written down as

$$R_z = \frac{Z_0}{k} \frac{1 + \left(1 + \frac{\langle h_t \rangle Z_0}{k}\right) \frac{\delta}{Z_0}}{\frac{\langle h_t \rangle Z_0}{k} + \left(1 + \frac{\langle h_t \rangle Z_0}{k}\right) \frac{\delta}{Z_0}}, \tag{5.63}$$

$$R_t = \frac{\sqrt{\alpha T_0}}{k} \frac{1 + \left(1 + \frac{\langle h_z \rangle \sqrt{\alpha T_0}}{k}\right) \frac{\delta}{\sqrt{\alpha T_0}}}{\frac{\langle h_z \rangle \sqrt{\alpha T_0}}{k} + \left(1 + \frac{\langle h_z \rangle \sqrt{\alpha T_0}}{k}\right) \frac{\delta}{\sqrt{\alpha T_0}}}. \tag{5.64}$$

In the case of absence of the fluctuations ($R_z = R_t = 0$), (5.62) takes a form characteristic for the standard three-part heat transfer

$$\frac{1}{U_m} = \frac{\delta}{k} + \frac{1}{\langle h_z \rangle} + \frac{1}{\langle h_t \rangle}. \tag{5.65}$$

It follows from relations (5.63) and (5.64) at $\delta \to \infty$ that

$$R_z = \frac{Z_0}{k}, \quad R_t = \frac{\sqrt{\alpha T_0}}{k}. \tag{5.66}$$

However, in this case the determining role in the general chain (5.62) is played by the thermal resistance of a body

$$\frac{1}{U_m} = \frac{\delta}{k} \to \infty. \tag{5.67}$$

At $\delta \to 0$ there are only crisscross terms left in relations (5.63) and (5.64):

$$R_z = \frac{1}{\langle h_t \rangle}, \quad R_t = \frac{1}{\langle h_z \rangle}. \tag{5.68}$$

Then, (5.62) can be written down as

$$\frac{1}{U_m} = \frac{\delta}{k} + \frac{2}{\langle h_z \rangle} + \frac{2}{\langle h_t \rangle}. \tag{5.69}$$

It is obvious from a comparison of expressions (5.65) and (5.69), that the limiting case of the zero-wall thickness is characterized by double decrease in the heat transfer intensity on each of the sides. This result can be formally treated as switching off a passive part of the full period of heat transfer. This implies a rather interesting conclusion: even at absence of a solid wall, two perturbed fluid flows remain in the thermal conjugation. This novel physical effect can exhibit itself for a distinctly expressed periodic structure of heat transfer for each of the sides. For example: on the left-hand side of a plate, a dropwise condensation (spatial periodicity) occurs, while on right-hand side a nucleate boiling (temporal periodicity) takes place.

A saliently expressed case of an obviously overestimated thermal effect of a wall is considered above. In a reality, the degree of the conjugation of three-part chains of heat transfer will be certainly far less expressed in a quantitative sense. In doing so, the author pursued a simple purpose to outline qualitative features of the problem of conjugate convective–conductive heat transfer in a general case characteristic for real heat exchangers [6–8].

References

1. Carslaw HS, Jaeger JC (1992) Conduction of Heat in Solids. Clarendon Press, London, Oxford.
2. Baehr HD, Stephan J (1998) Heat and Mass Transfer. Springer, Berlin Heidelberg New York.
3. Sagan H (1989) Boundary and Eigenvalue Problems in Mathematical Physics. Dover Publications, New York.
4. Watson GN (1995) A Treatise on the Theory of Bessel Functions. Cambridge University Press.
5. Kays WM, Crawford ME, Weigand B (2004) Convective Heat and Mass Transfer. Mc Graw Hill, New York.
6. Roetzel W, Spang B (2002) Berechnung von Wärmeübertragern, Wärmedurchgang, Überschlägige Wärmedurchgangskoeffizienten. VDI – Wärmeatlas, Ca – Cc. Springer.
7. Roetzel W, Xuan Y (1999) Dynamic Behaviour of Heat Exchangers. WIT Press/Computational Mechanics Publications, Southampton.
8. Piccolo A, Pistone G (2006) Estimation of heat transfer coefficients in oscillating flows: The thermoacoustic case. Int J Heat Mass Transf 49: 1631–1642.

6

Step and Nonperiodic Oscillations of the Heat Transfer Intensity

6.1 Asymmetric Step Oscillations

Three basic characteristic functions $\psi(\xi)$ investigated in Chap. 3 (harmonic, inverse harmonic, and symmetric stepwise) are single parametrical in the sense that their form is unequivocally characterized by the relative amplitude of oscillations b. An essential expansion of the class of the periodic conjugate problems can be achieved via prescribing an asymmetric step function $\psi(\xi)$ (Fig. 6.1)

$$h_\mathrm{m} = \langle h \rangle \varepsilon = h_+ s\varepsilon. \tag{6.1}$$

The relative amplitudes of oscillations for both the active ($\psi = 1+b$) and passive ($\psi = 1-a$) periods, as well as the proportion of their durations are connected among themselves by a normalizing relation

$$bs = a(1-s). \tag{6.2}$$

Here the value of s (parameter of asymmetry) is defined as a ratio of the duration of the active period τ_+ to that of the full period of oscillations τ_0:

$$s = \frac{\tau_+}{\tau_0}. \tag{6.3}$$

Since the asymmetric (two-parametrical) step function determined by relations (6.1) and (6.2) cannot be presented as the Fourier series [1], the method of orthogonalization (used in Sect. 3.5 at the analysis of the symmetric single-parametrical case) cannot be applied for an analysis of this function. A substitution of (6.1) and (6.2) in the computational algorithm (4.20) can yield

$$\varepsilon = \frac{a(1-a) + s\left[(1-a)^2 + \chi\right]}{a + s(1-2a+\chi)}, \tag{6.4}$$

where $\chi = H/\langle \bar{h} \rangle$ is the PTE [(4.17)]. Assuming $s = 1/2$, one can easily obtain from (6.4) a relation for the case of symmetric stepwise oscillations considered in Sect. 3.5.

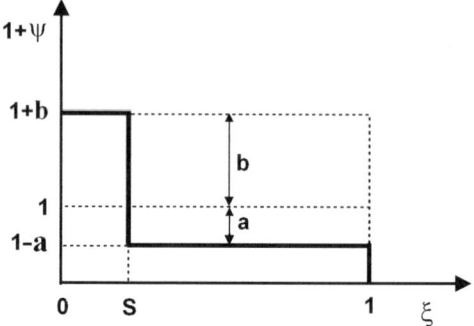

Fig. 6.1. Asymmetric step pulsations of the THTC

$$\varepsilon = \frac{1 - a^2 + \chi}{1 + \chi}. \tag{6.5}$$

The contrast between the extremely simple formula (6.5) and rather complex equations (3.12) and (3.26) is striking. Calculations for the case of a time-dependent problem and a semi-infinite body show that the relative deviation of (6.5) from the exact solution can reach, as a maximum, only 8%. Let us try now to obtain a more detailed description of the given two-parametrical case. For this purpose, we shall carry out a separate analysis of the problems of unsteady heat transfer for a semi-infinite body separately for each of the active and passive periods. In doing so, we shall consider a problem of purely temporal periodicity of heat transfer, whose solutions are well documented [2].

Semi-infinite body. Let us consider a case of a semi-infinite body $\delta \to \infty$ with a uniform initial temperature distribution $\vartheta = \vartheta_{\max}$. At the initial moment of time, heat transfer with an ambience at a constant heat transfer coefficient $h_+ = \mathrm{const}$ is switched on. The unsteady temperature field in the body can be described by a solution [2]

$$\frac{\vartheta}{\vartheta_{\delta \max}} = \mathrm{erf}\left(\frac{X}{2\sqrt{\alpha\tau}}\right) + \exp\left[\frac{h_+ X}{k} + \left(\frac{h_+\sqrt{\alpha\tau}}{k}\right)^2\right] \mathrm{erfc}\left(\frac{X}{2\sqrt{\alpha\tau}} + \frac{h_+\sqrt{\alpha\tau}}{k}\right). \tag{6.6}$$

The temperature of the body's surface varies according to the law

$$\frac{\vartheta_\delta}{\vartheta_{\delta \max}} = \exp\left(\tilde{h}_+^2 \tilde{\tau}\right) \mathrm{erfc}\left(\tilde{h}_+ \sqrt{\tilde{\tau}}\right). \tag{6.7}$$

Here

$$\tilde{h}_+ = \frac{h_+ \sqrt{\alpha\tau_+}}{k} \tag{6.8}$$

is the Biot number based on the period of cooling τ_+ (i.e., the active period of heat transfer),

6.1 Asymmetric Step Oscillations

$$\tilde{\tau} = \frac{\tau}{\tau_+} \tag{6.9}$$

is dimensionless time. In the end of the active period of heat transfer (or the period of cooling) the temperature will reach a minimal value equal to

$$\frac{\vartheta_{\delta \min}}{\vartheta_{\delta \max}} = \exp\left(\tilde{h}_+^2\right) \operatorname{erfc}\left(\tilde{h}_+\right). \tag{6.10}$$

Then the passive period is switched on, during which the surface of the body is adiabatic, and its temperature should grow due to the heat flux from inside of the body. From the reasons of preserving the physical dimensions in both parts of the equation, the law of the surface temperature rise looks like

$$\vartheta_\delta = \vartheta_{\delta \min} + \frac{2}{\sqrt{\pi}} \frac{\langle q \rangle \sqrt{\alpha \tau}}{k}, \tag{6.11}$$

where $\langle q \rangle$ is the average heat flux on the surface of the body for the full period of time. It is interesting to point out that expression (6.11) at $\vartheta_{\min} = 0$ coincides with the known solution [2] for the case of heating of a semi-infinite body due to a constant heat flux supply to its boundary. Splicing together the active and passive periods, one can obtain a solution for the case of asymmetric stepwise oscillations. The ATHTC can be calculated then from an obvious correlation

$$\langle h \rangle = h_+ s. \tag{6.12}$$

An average temperature difference $\langle \vartheta_\delta \rangle$ for the full heat transfer period is defined via averaging two correspondent average values for active $\langle \vartheta_{\delta+} \rangle$ and passive $\langle \vartheta_{\delta-} \rangle$ periods:

$$\langle \vartheta_\delta \rangle = \langle \vartheta_{\delta+} \rangle s + \langle \vartheta_{\delta-} \rangle (1 - s). \tag{6.13}$$

The average value of the heat flux and the average temperature difference are interrelated as

$$\langle q \rangle = h_m \langle \vartheta_\delta \rangle. \tag{6.14}$$

From here, an equation for the EHTC can be derived:

$$h_m = \langle h \rangle \varepsilon = h_0 s \varepsilon. \tag{6.15}$$

Substituting (6.1) at $a = 1$ into the boundary condition (BC) (2.29) and performing the procedure of averaging, one can obtain a relation for the factor of conjugation (FC)

$$\varepsilon = \frac{\langle \vartheta_{\delta+} \rangle}{\langle \vartheta_\delta \rangle}. \tag{6.16}$$

Omitting simple intermediate transformations, we shall write down the final expression for the FC

$$\varepsilon = \frac{1}{s + (1 - s) F_1}. \tag{6.17}$$

Here F_1 is the function determined by a ratio

$$F_1 = \frac{A^2 \left[2 - A + (1 + A^2)^{1/2}\right]}{2\left[(1 + A^2)^{3/2} - 1 - A^3\right]}, \quad (6.18)$$

where

$$A = \frac{\pi}{\sqrt{2}} \tilde{h}_+.$$

It is expedient to deduce an approximate expression for the function F_1, which practically coincides with (6.18) (with the maximal inaccuracy up to 0.3% over the whole range of variation of the parameter A) and is more convenient for calculations

$$F_1 = \frac{1 + 0.672A + 0.183A^2}{1 + 0.315A}. \quad (6.19)$$

Introducing a temperature scale by the ratio

$$\vartheta_* = \frac{\langle q \rangle}{h_+}, \quad (6.20)$$

one can find out from relations (6.5)–(6.9) the minimal $\vartheta_{\delta\,\min}$ and the maximal $\vartheta_{\delta\,\max}$ (over the period) temperature differences

$$\frac{\vartheta_{\delta\,\min}}{\vartheta_*} = \frac{3F_1 F_2}{(2 + F_2)s}, \quad (6.21)$$

$$\frac{\vartheta_{\delta\,\max}}{\vartheta_*} = \frac{3F_1}{(2 + F_2)s}, \quad (6.22)$$

$$F_2 = (1 + A^2)^{1/2} - A. \quad (6.23)$$

Advantages of the above-mentioned description of the two-parametrical stepwise case consist in an opportunity to determine not only the FC, but also the law of the temperature variation over the heat transfer surfaces in time.

Symmetric step function. The solution (6.17)–(6.19) at $s = 1/2$ describes the case of the symmetric step function for the time-dependent problem at $b = 1$ and agrees (with the maximal relative error less than 3%) with the exact solution (3.56) obtained in Chap. 3. This encouraging circumstance serves as a kind of a test allowing validating the correctness of the computational algorithm for arbitrary values of the asymmetry parameter s.

Delta-like step function. Let us fix up a value of the THTC during the active period of heat transfer $h_+ = $ const and tend the parameter of asymmetry to zero $s \to 0$. In doing so, it can be obtained from (6.17)–(6.19)

$$\varepsilon = \frac{2\left[(1 + A^2)^{3/2} - 1 - A^3\right]}{A^2 \left[2 - A + (1 + A^2)^{1/2}\right]}. \quad (6.24)$$

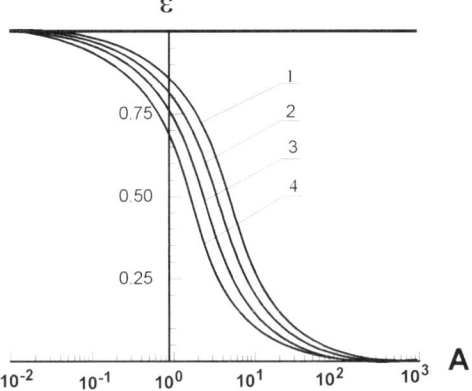

Fig. 6.2. Asymmetric step pulsations of the THTC. Dependence of the FC on the Biot number: (1) $s = 0.65$, (2) $s = 0.5$, (3) $s = 0.3$, (4) $s = 0$

It follows from relation (6.24), that the FC is determined only by the value of $\tilde{h}_+$ (Fig. 6.2). Now, if one fixes up a value of the average (for the full period) THTC $\langle h \rangle = \text{const}$, then (6.17)–(6.19) at $s \to 0$ describe the asymptotical case of delta-like oscillations of the THTC

$$\varepsilon = \frac{2\sqrt{2}}{\pi} \frac{\sqrt{s}}{\langle \tilde{h} \rangle}. \tag{6.25}$$

From (6.25), one can obtain an expression for the EHTC

$$\alpha_m = \frac{2\sqrt{2}}{\pi} \frac{\sqrt{kc\rho\tau_+}}{\tau_0}. \tag{6.26}$$

It is interesting to compare (6.26) with the delta-like asymptotical solution for the inverse harmonic type of oscillations of the THTC following from (4.22)

$$h_m = k^{1/2} \left(\frac{\langle \tilde{h} \rangle h_{\min}}{2\alpha\tau_0} \right)^{1/4}. \tag{6.27}$$

First of all, relations (6.26) and (6.27) are interesting because of their obviously nontrivial form. They show that for different laws of oscillations of the THTC the delta-like transition results in completely different results.

Thin wall for the TBC $q_0 = \text{const}$. For an analysis of the two-parametrical law of oscillations of the THTC in the case of a thermally thin wall $\delta \to 0$ at the TBC $q_0 = \text{const}$, it is possible to use the equation of heat conduction averaged over the cross-sectional coordinate with THTC presented as a source term [3]

$$c\rho\delta \frac{d\vartheta}{d\tau} = q - h\vartheta. \tag{6.28}$$

Performing a procedure of splicing together of the active and passive periods (similarly to the considered above case) and omitting simple derivations, one can obtain an equation for the FC

$$\varepsilon^{-1} = 1 + (1-s)^2 \left(C \coth(C) - 1 \right), \qquad (6.29)$$

where

$$C = \frac{1}{2} \frac{h_+ \tau_+}{c\rho\delta}. \qquad (6.30)$$

As the solution (6.29) is used here as the asymptotic case of a infinitely thin wall, it is necessary to execute in this solution a limiting transition for $\delta \to 0$, i.e., for $C \to \infty$. From here, one can obtain a limiting equation for the FC

$$\varepsilon_0 = \frac{1}{(1-s)^2 C}. \qquad (6.31)$$

Splicing of the asymptotical solutions. An analysis of the dependences $\varepsilon(\delta)$ obtained for the two-parametrical case (Fig. 3.12) shows that these dependencies with a good accuracy can be approximated via a simple two-zone splicing of the correspondent asymptotical solutions for a semi-infinite body and a thin wall. As applied to the two-parametrical case considered here, the solution for an arbitrary wall thickness can be written down as

$$\left. \begin{array}{l} 0 \leq \tilde{\delta} \leq \tilde{\delta}_0 : \varepsilon = \varepsilon_0, \\ \tilde{\delta}_0 \leq \tilde{\delta} < \infty : \varepsilon = \varepsilon_\infty. \end{array} \right\} \qquad (6.32)$$

Here

$$\tilde{\delta}_0 = \frac{3(1-s)^2 \tilde{h}_+}{6 + 2\sqrt{\pi}(1-s)\tilde{h}_+}, \qquad (6.33)$$

$\tilde{\delta} = \delta/\sqrt{\alpha\tau_+}, \varepsilon_\infty, \varepsilon_0$ are limiting relations for the FC determined by (6.17)–(6.19) and (6.31), respectively.

Arbitrary amplitude of oscillations. Relations (6.17)–(6.19) and (6.31) were received for a limiting case of the maximal amplitude of oscillations of the THTC described by the step function at $a = 1$. This case physically corresponds to the adiabatic passive period of heat transfer. For a transition to the general case of an arbitrary amplitude $0 \leq a \leq 1$, an approach based on the analogy to the reduced form of the solution for the symmetric step function [(3.76) and (3.77)] can be used

$$\varepsilon = \varepsilon_{\min} + (1 - \varepsilon_{\min}) \varepsilon_*. \qquad (6.34)$$

Here ε_* is the reference value of the FC computed from relations (6.17)–(6.19) (for a semi-infinite body) or (6.31) (for a thin wall); $\varepsilon_{\min}$ is the minimal value of the FC determined by a ratio

$$\varepsilon_{\min} = \frac{(1-a)\left[a + (1-a)s\right]}{a - (1-2a)s}, \qquad (6.35)$$

a is the relative amplitude of oscillations for the passive period included in the ratio (6.2). Assuming $a = 1$ in (6.34) and (6.35), one can come to the case of the adiabatic passive period $\varepsilon_{\min} = 0$ considered above. In the conclusion to this section, it is necessary to point out that we considered above only the case of the TBC $q_0 =$ const. Unfortunately, it is not possible to obtain similar simple formulas for the alternative TBC of $\vartheta_0 =$ const.

6.2 Nonperiodic Oscillations

It was assumed everywhere above that we have dealt exclusively with periodic functions represented as the Fourier series. This certainly relates in the full extent also to the computational algorithm (4.20). Therefore, the extension of the method developed in the present work for the class of nonperiodic functions falls outside of the scope of the problems considered in this book. However, the aforementioned filtration property of the algorithm (4.20) smoothing peaks and high-frequency components of oscillations of the THTC has inspired us (generally speaking, without any substantiation) to try to intrude in the area of nonperiodic oscillations. In this case, additionally to the fundamental problems, a technical problem of averaging immediately arises. At the analysis of periodic functions, the issue of averaging does not arise at all, as it can be resolved automatically due to the representation of the THTC as a Fourier series [1]. A transition to the nonperiodic functions demands in each case a separate definition of the "average over the period" value of the THTC. Only after doing that, it is possible to compare results of the solutions of the periodic and nonperiodic problems. For the sake of simplicity, we shall write down everywhere below the function $\psi(t)$ dependent only on time. It would be formally possible also to write down here again a function of the coordinate of a progressive wave $\psi(\xi)$, however, in this case this does not make any difference. Besides, in the applications associated with oscillations, as a rule, one always deals with temporal oscillations. The author realizes that the transition from the strictly periodic (the Fourier series) functions to the "conditionally periodic" functions does not have any other substantiation, except for only intuition.

Functions of the kind $\sin(1/t)$. A function of this kind (Fig. 6.3) can be considered conditionally periodic over the interval of time $t_2 - t_1$, if one presents it as

$$\psi = \sin\left(\frac{1}{t}\right) - \frac{\int_{t_1}^{t_2} \sin(1/t)\, dt}{t_2 - t_1}. \tag{6.36}$$

Expression (6.36) can be transformed in view of the integral

$$\int_0^T \sin\left(\frac{1}{t}\right) dt = T\sin\left(\frac{1}{T}\right) - \mathrm{Ci}\left(\frac{1}{T}\right), \tag{6.37}$$

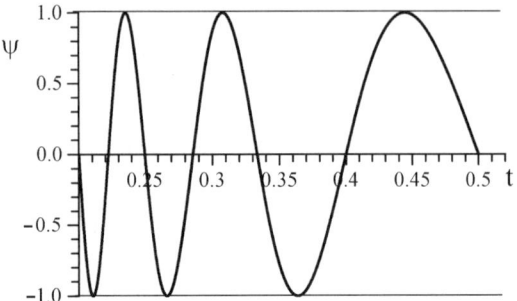

Fig. 6.3. Nonperiodic pulsations of the THTC of the kind $\sin(1/t)$

where Ci(x) is the integrated cosine function [4].[1] Moving along the T-axis from infinity to zero, we one can find out that the function Ci$(1/T)$ has the first zero at $T \cong 0.295$. Therefore, we accept the coordinate of the right-hand side border of trimming equal to $t_2 = 0.295$. Reducing the variable T within the interval $0 \leq T \leq 0.295$, one can consecutively pass zero points of the function Ci$(1/T)$ with the constantly smaller step on the axis T. At last, at the point $T = 0$, the function Ci$(1/T)$ becomes uncertainty. Therefore, for a particular analysis, it is necessary to set a certain value of the left-hand side coordinate of trimming $t_1 > 0$. As a criterion for the choice of this coordinate, the number of zero points n within the chosen interval (for example, $n = 10$) can serve. If limits of integration in (6.36) coincide with the correspondent zero points of the function Ci$(1/T)$, the integral in the right-hand side of (6.36) drops out, and one can obtain as a result

$$\psi = \sin\left(\frac{1}{t}\right). \tag{6.38}$$

Taking consecutively zero points of the function Ci$(1/T)$ within the chosen interval, one can thus shift (reduce) the value of t_2 down to its coincidence with the value t_1. It gives at our disposal a final set of the conditionally periodic functions in the form of (6.36). Substituting the function $\psi = \sin(1/t)$ into the algorithm (4.20), one can obtain the final solution as a dependence of the FC on the PTE. Computed dependences of the function $\varepsilon(\chi)$ agree quite fairly with correspondent dependences for the harmonic law of oscillations. Thus, in spite of the strong distinctions between the actual functions $\psi(t)$, the symmetrically smooth character of oscillations causes prevailing influence on the factor of conjugation, while the "floating" period of the nonperiodic function, as well as its deformation over this period, practically do not affect the value of the FC.
Functions of the kind $\sin(\sqrt{t})$. This function (Fig. 6.4) is conditionally periodic over the interval of time $t_2 - t_1$, $0 - t_1$. We omit here the insignificant intermediate

---

[1] One has to point out that the transition $t \Rightarrow T$ has been made here to avoid a misunderstanding at changing the notation for time: t designates here an integrand variable.

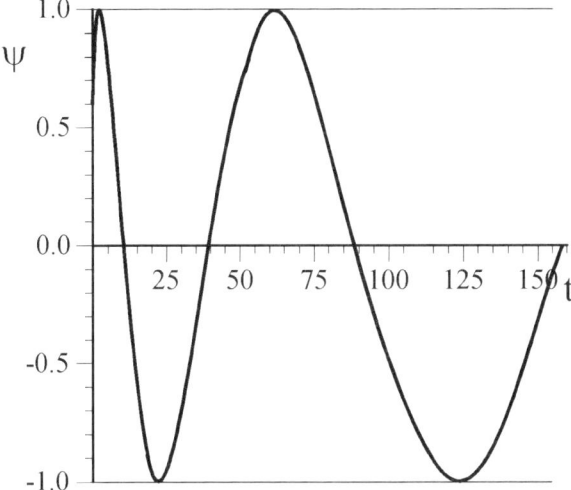

Fig. 6.4. Nonperiodic pulsations of the THTC of the kind $\sin\left(\sqrt{t}\right)$

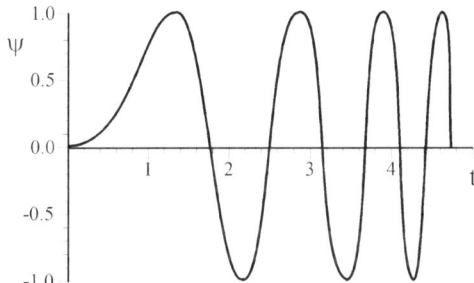

Fig. 6.5. Nonperiodic pulsations of the THTC of the kind $\sin\left(t^2\right)$

derivations completely similar to those done above. Computed dependences of the function $\varepsilon(h)$ also agree well with the calculations for the harmonic law of oscillations.

Functions of the kind $\sin\left(t^2\right)$. A difference in this case in comparison with the previous one consists in a shift of the range of definition of the function $\psi(t)$ to the right along the T-axis (Fig. 6.5). A good agreement with the harmonic law also takes place here.

Functions of the kind $\sin(\ln t)$. The range of definition of this function is $1 \leq t < \infty$. Calculations show only a rather weak qualitative similarity with the form of the dependence $\varepsilon(\chi)$ for the harmonic law of oscillations. One can not talk in this case about any quantitative comparison. Thus, such a strong nonperiodical distortion of oscillations engenders results strongly different from the correspondent "periodic solutions." So, the intrusion into the area of nonperiodic oscillations undertaken actually without any good substantiation has led us to inconsistent results. For the first three considered functions $\psi(t)$, the smoothing property of the computational algorithm effectively exhibits itself, and the nonperiodic solutions for the FC differ from the periodic ones only a little. In other words, the deformations imposed by nonperiodicity on the primarily smooth (periodic) oscillations, for the considered cases are practically not reflected in the final dependence for the function $\varepsilon(\chi)$. These encouraging results, however, are actually brought to nothing by the solution for the fourth nonperiodic function $\psi(t)$, which completely drops out of the overall picture. As it is deemed, the generalization of the method developed in the present book for the case of nonperiodic and stochastic oscillations of thermohydraulic parameters is basically possible via use of the Fourier transform technique and wavelet-analysis [5]. However, these methods lie far away from the scope of the present book.

References

1. Stein EM, Shakarchi R (2003) Fourier Analysis: An Introduction. Princeton University Press, Princeton.
2. Carslaw HS, Jaeger JC (1992) Conduction of Heat in Solids. Clarendon Press, London, Oxford.
3. Baehr HD, Stephan K (1998) Heat and Mass Transfer. Springer, Berlin Heidelberg New York.
4. Abramovitz M, Stegun IA (1974) Handbook of Mathematical Functions with Formulas, Graphs, and Mathematical Tables. Dover Publications, New York.
5. Gasquet C, Witomski P, Ryan V (1998) Fourier Analysis and Applications: Filtering, Numerical Computation, Wavelets. Springer, Berlin Heidelberg New York.

7
Practical Applications of the Theory

7.1 Model Experiment

In order to illustrate the influence of thermophysical properties of a solid body on the experimental heat transfer coefficient (EHTC) under conditions where heat transfer intensity is subjected to periodic oscillations, a special model experiment has been designed and carried out. Its purpose was to determine a dependence of the function $\varepsilon\left(\left\langle\tilde{h}\right\rangle\right)$ for a semi-infinite body under conditions of a time-dependent problem. This dependence has been theoretically computed above [(3.56)] and shown in Fig. 3.10. The basic element of the experimental rig (Fig. 7.1) was a long brass electrically heated rod (1) thermally insulated on its lateral cylindrical surface, with the end face being periodically washed with a colder water jet from the nozzles (2) of various diameters. It allowed modeling a problem of oscillations of the heat transfer intensity in time according to the symmetric step law with the amplitude close to maximal: $b \approx 1$. Such a simple experiment allowed determining the required dependence $\varepsilon\left(\left\langle\tilde{h}\right\rangle\right)$ via direct measurements, at different nozzle diameters, of all the necessary parameters such as the Biot number, ATHTC and the EHTC (see Table 7.1). The temperature curves for the water-cooled surface for different values of the cooling period and the nozzle diameters are shown in Fig. 7.2. A qualitative agreement with the correspondent theoretical curves obtained in Chap. 3 (Fig. 3.6) is evident. As one can see from Fig. 7.3, the experimental and theoretical dependences $\varepsilon\left(\left\langle\tilde{h}\right\rangle\right)$ agree well among themselves.

The model experiment evidently illustrates the method of the analysis of the periodic conjugate heat transfer processes developed in the present book. The THTC is preset by the external water jet cooling of the side surface and, thus, it is hydrodynamically determined. The EHTC is being found from the experimentally realized model of a boundary problem for the equation of heat conduction with the TBC of the third kind. It is important to note that in doing so the significant quantitative influence of the thermal effect of a solid

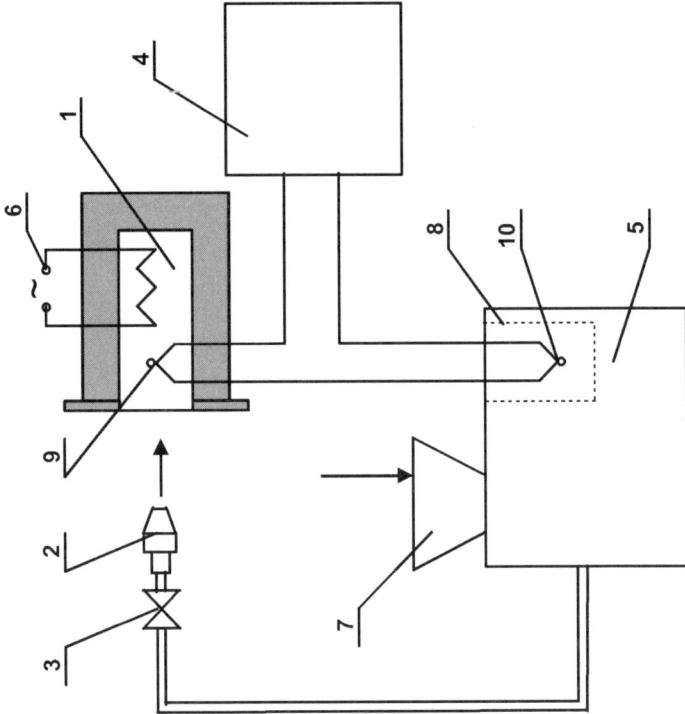

Fig. 7.1. Model experiment: (1) brass rod of the length 300 mm and diameter 8 mm, (2) nozzle (of the diameter 2, 3, 5, 6 mm), (3) electromagnetic valve, (4) mirror-galvanometer oscillograph, (5) thermostat, (6) nickel-chromium heater, (7) sink arrangement, (8) Dewar flask, (9) thermocouple in the brass rod, (10) thermocouple in the thermostat

body has been achieved, with the smallest values of the factor of conjugation (FC) reaching $\varepsilon \approx 0.13$.

7.2 Dropwise Condensation

As it is known [1], heat transfer at dropwise condensation is characterized by spatial nonuniformity caused by an intermittent location on a solid surface of the large droplets and a thin film of a condensed liquid. An essential influence of heat conduction in the wall on the EHTC has been revealed, in particular, in works [1–6]. Figure 7.4 shows a schematic of an elementary cell involved in the condensation process. As one can see from this figure, it is possible with a good degree of accuracy to approximately describe the process of the dropwise condensation by the correspondent spatial step law of oscillations of the THTC $h(Z)$ with the amplitude close to maximal ($b \approx 1$). For a quantitative calculation of the thermal effect of a solid body within the framework of the developed method, it is necessary at first to determine the ATHTC. It has been

Table 7.1. Parameters of the model experiment

(a) $d_0 = 2\,\text{mm}$		
$\tau_0(s)$	$\langle \tilde{h} \rangle$	ε
1.6	0.382	0.788
5	0.675	0.686
10	0.952	0.582
30	1.65	0.438
60	2.34	0.340
120	3.3	0.265
240	4.68	0.202
480	6.62	0.139
(b) $d_0 = 3\,\text{mm}$		
1.6	0.256	0.813
5	0.456	0.756
10	0.622	0.655
30	1.11	0.523
60	1.56	0.461
120	2.22	0.331
240	3.13	0.26
480	4.43	0.199
(c) $d_0 = 5\,\text{mm}$		
1.6	0.162	0.887
5	0.286	0.846
10	0.404	0.784
30	0.700	0.670
60	0.989	0.560
120	1.405	0.462
240	1.985	0.382
480	2.800	0.276
(d) $d_0 = 6\,\text{mm}$		
1.6	0.105	0.895
5	0.186	0.90
10	0.264	0.865
30	0.455	0.725
60	0.643	0.620
120	0.91	0.582
240	1.29	0.507
480	1.82	0.402

shown above that values of $\langle h \rangle$ and h_m, are equal to each other at $k \to \infty$. This effectively means that it is possible to consider the ATHTC approximately equal to the EHTC measured in experiments with a material possessing very high thermal conductivity (for example, for a case of a copper wall). The spatial scale of periodicity can be assumed being approximately equal to the maximal size of a large droplet $Z_0 = \beta r_0$. Then we can calculate the Biot number $\left(\langle \tilde{h} \rangle = \langle h \rangle Z_0/k \right)$ and further fulfill comparisons of the experimental

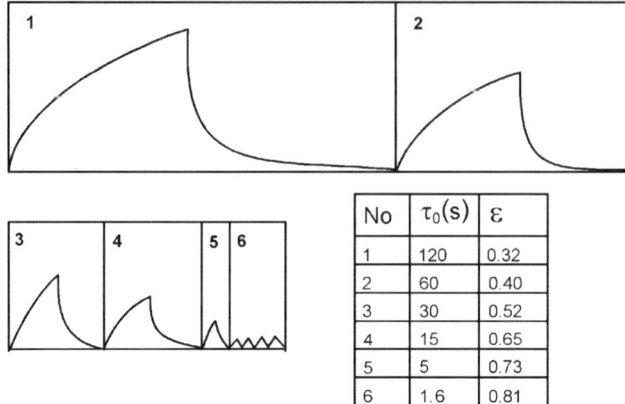

Fig. 7.2. Dependence of the temperature of the water-cooled surface on time: (1) $\tau_0 = 120\,\text{s}$, (2) $\tau_0 = 60\,\text{s}$, (3) $\tau_0 = 30\,\text{s}$, (4) $\tau_0 = 15\,\text{s}$, (5) $\tau_0 = 5\,\text{s}$, (6) $\tau_0 = 1.6\,\text{s}$

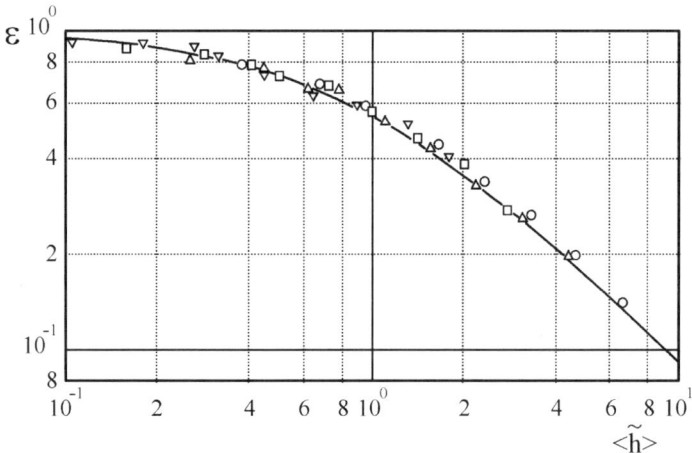

Fig. 7.3. Experimental (*points*) and theoretical (*line*) dependencies of the FC on the Biot number: *open circle*, $d_0 = 2\,\text{mm}$; *open triangle*, $d_0 = 3\,\text{mm}$; *open square*, $d_0 = 5\,\text{mm}$; *inverted triangle*, $d_0 = 6\,\text{mm}$

and computational data characterizing the influence of the heat conduction in a body on the EHTC. As shown in Fig. 7.5 in semilogarithmic coordinates, if the numerical constant takes the value of $\beta \approx 4.8$ the theoretical model qualitatively truly and quantitatively fairly well reflects the physical tendency of the thermal effect of a body on the average heat transfer.

A special experiment designed for the heat transfer measurements at dropwise condensation was carried out in [7] for various values of the spatial lengthscale r_0 and with the same material of the walls (gold). In this case, the maximal size of a droplet was adjusted by varying the centrifugal force in the rotating system. It was obtained that, at increase in the size r_0, the value of the EHTC decreased. As follows from Fig. 7.6, an application of the above-developed computational

7.2 Dropwise Condensation   125

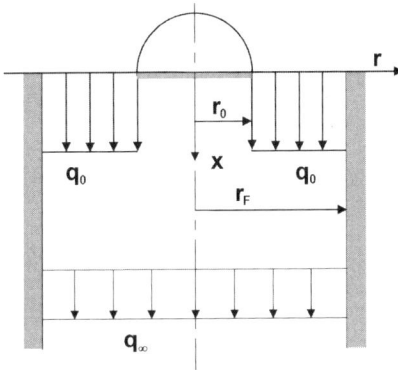

Fig. 7.4. Schematic of the process of dropwise condensation

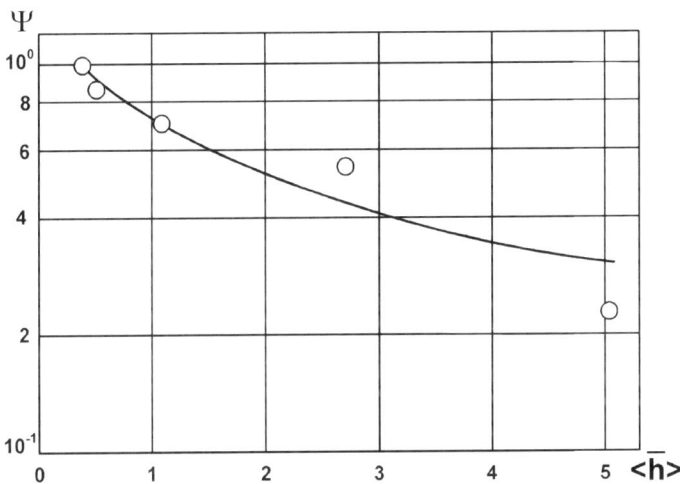

Fig. 7.5. Effect of heat conduction in the body on the EHTC at dropwise condensation

technique to experiments [1] leads to a satisfactory agreement between the theory and experiments. Standing on the positions of the model developed above, it is possible to treat the experimentally revealed facts of the influence of the thermal conductivity of a wall and the effect of the droplet radius on the average heat transfer at dropwise condensation as the particular cases exhibiting the dependence of the FC on the Biot number $\langle \bar{h} \rangle = \langle h \rangle Z_0/k$.

The theoretical model of heat transfer at dropwise condensation was considered earlier [1]. The reasoning of the author of the work [1] can be illustrated with the help of Fig. 7.4. It was assumed that the removal of heat occurs through a thin ring-like film of the liquid surrounding a large droplet, which shields (or, in other words, protects) the heat transfer surface. In doing so, the boundary condition $q = \text{const}$ was accepted for the nonshielded (ring-like) surface. This effectively means a statement of a stationary boundary problem for the heat conduction equation with the TBC of the second kind. However, in addition to

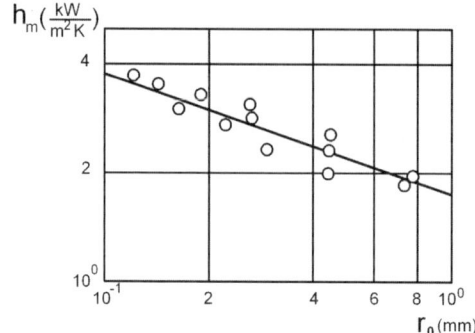

Fig. 7.6. Effect of the maximal lengthscale of a droplet on the EHTC at dropwise condensation

it, the TBC of the third kind was used in the work [1] at the transition to the determination of the EHTC. It is clear that a simultaneous use of these two TBC results in the fact that the problem becomes overdefined from the mathematical point of view. As an indirect evidence of this, one can judge such a fact that the final dependence presented in the work [1] looks like $h_m = f(k/k_f)$. At the same time, a correct dependence for the case of the TBC of the third kind should look like $h_m = f\langle \bar{h} \rangle$. A similar approach has been applied further in the work [6] in order to take into account the dependence of the average intensity of heat transfer on the wall thickness in case of dropwise condensation. However, in view of the said above, these results should be considered equally disputable, like the results of the work [1].

The problem of averaging the heat transfer coefficient (HTC) at dropwise condensation was apparently for the first time discussed in the works [8,9]. The authors of the works [8,9] based their reasoning on the understanding that, for a determination of the ATHTC, it is necessary in the beginning to calculate the THTC as a quotient from a division of a local value of the heat flux by the temperature difference (see Fig. 1.1). Then, averaging of the THTC can provide the required value of $\langle h \rangle$, which, in accordance with the fundamental hypothesis of the present research, is practically independent of the properties of the wall. As to the experiments [1–6], they provided measurements of the namely traditional value of h_m, i.e., a quotient from a division of the correspondent parameters (already averaged during the experiments). As we believe, namely for this reason an essential influence of the thermophysical properties of the wall on the EHTC was pointed out in the aforementioned works.

7.3 Nucleate Boiling

7.3.1 Theory of Labuntsov

As it is known [10,11], the process of nucleate boiling under conditions of natural convection is characterized by periodic oscillations of parameters both in time (origin, growth and separation of vapor bubbles), and lengthwise, i.e., along the

7.3 Nucleate Boiling

heat transfer surface (due to existence of the spatially fixed nucleation sites). As far as it is known to the author, historically the first theoretical model of heat transfer at nucleate boiling has been proposed in 1963 by Labuntsov (see [11]). In this work, on the basis of an analogy with a near-wall turbulent flow, a concept of a "friction velocity" was introduced

$$u_* = \frac{R\dot{R}}{L}. \tag{7.1}$$

Here R is a radius of a vapor bubble growing on the heated wall with the growth rate equal to $\dot{R} = dR/d\tau$,

$$L = 10^4 R_* \tag{7.2}$$

is the distance between the boiling nucleation sites (or, in other words, bubble-producing sites),

$$R_* = 2\frac{\sigma T_s}{h_{fg}\rho_g \vartheta} \tag{7.3}$$

is the minimal possible radius of a germinating vapor bubble (vapor nucleus). A theoretical dependence of the radius of a vapor bubble on time has been found also in the work [11]

$$R = \sqrt{12\frac{k_f \vartheta \tau}{h_{fg}\rho_g}}. \tag{7.4}$$

Assuming that the main part of the thermal resistance at nucleate boiling is concentrated in the thin near-wall layer of a thickness

$$\delta_f = 12\frac{\nu_f}{u_*}, \tag{7.5}$$

Labuntsov wrote down the HTC at nucleate boiling as

$$h = \frac{k_f}{\delta_f}. \tag{7.6}$$

From (7.2)–(7.6), known formula of Labuntsov for heat transfer at nucleate boiling follows

$$q = 10^{-3} \frac{k_f^2 \vartheta^3}{\nu_f \sigma T_s}. \tag{7.7}$$

Equation (7.7), which is based on the understanding of the nucleate boiling as a specific case of near-wall turbulence, was obtained grounding on the analysis of microcharacteristics of the process such as sizes and growth rates of individual vapor nuclei. Since the middle of 60th years of the last century, in the publications devoted to the experimental investigations of nucleate boiling, filmograms of the near-wall areas of a boiling liquid have began to appear. These data testified that the mode of boiling with individual vapor bubbles exists only at a rather small superheating of a liquid. For the developed boiling regime, the entire process becomes essentially complicated. It is possible to distinguish here a nonstationary liquid film adjoining to the heated surface and vapor conglomerates existing in the liquid and connected to the wall with vapor columns.

After the visual information about nucleate boiling has appeared, Labuntsov in 1972 (see [11]) offered a new derivation of (7.7) based on an analysis of the macrocharacteristics of the process. He grounded his analysis on the classical solution of the Navier–Stokes equations for a fluid flow caused by harmonic oscillations of a wall in its own plane [12] (see Chap. 1, (1.17)–(1.19)). In this case, in the liquid, a near-wall layer (decelerated due to viscosity) can be distinguished of a thickness

$$\delta_f \sim \sqrt{\nu_f \tau_0}. \tag{7.8}$$

The period of oscillations was estimated in the work [11] on the basis of an analysis of physical dimensions and expressed in the following form

$$\tau_0 \sim \frac{L}{w_g}, \tag{7.9}$$

where

$$w_g = \frac{q}{h_{fg}\rho_g} \tag{7.10}$$

is the surface-averaged nucleation rate. Using the above-listed formulas in (7.6), one can come again to the previously derived (7.7). This remarkable property allowed Labuntsov to propose a hypothesis of a self-similarity of the nucleate boiling [11].

As far as it is known to the author, in spite of the long-term theoretical and experimental investigations of nucleate boiling, nobody has created so far a closed theory of this extremely complex process [13–17]. This conclusion is confirmed by extensive survey works [13–15], which just reproduced or modified the equations for the HTC calculation developed during 70th years of the last century. These equations have been grounded on the basis of an analysis of microcharacteristics of the process and, thus, they remain in the mainstream of the approach [11]. A rather interesting direction in the field of the boiling theory based on numerical modeling of this process [16] uses a series of initial assumptions, contains a significant number of numerical constants and is consequently still quite far from its final completion.

At last, attempts to bypass the basic difficulties connected to insufficient knowledge of the internal structure of the boiling process with the help of the formal mathematical methods borrowed from other areas of physics are believed to be unpromising. This relates, in particular, to the "fractal models" of heat transfer at nucleate boiling suggested in the work [17]. Concepts of a "fractal" and "fractal geometry" introduced in 1975 by Mandelbrot [18] relate to the irregular (chaotic) structures possessing a property of self-similarity. In the simplified terms it means that some small part of a fractal already contains in a compressed form the information on the entire fractal as a whole. Fractals indeed play an important role in the theory of nonlinear dynamic systems, where they allow, with the help of simple algorithms, investigating complex and nontrivial structures [19]. In view of the said above, it would be possible to expect from the work [17] some qualitatively new results in the modeling of nucleate boiling, like it, for example, did happen at the fractal analysis of chaotic structures [19]. However, in fact, the authors of the work [17] restricted themselves with a search of the formulas for the parameters of nucleate boiling (characteristic time of growth of a vapor bubble, thickness of a temperature boundary layer, nucleation site density, HTC, etc.) already available in the literature, and then the authors simply put certain concepts from the theory of fractals in correspondence with these equations.

To summarize, it is possible to agree with the capacious definition stated by the author of the work [10]: "...Heat transfer at boiling is always determined by simultaneous influence of numerous mechanisms controlling transfer of a substance...." These words can be understood as an expression of constrained pessimism concerning an opportunity of the solution of the problem of boiling in the foreseeable future. Here an analogy arises to the known problem of the theory of turbulence, which is known to be also rather far from the final completion [20]. Such a stand-point gets even the greater weight in view of the fact that the author of the work [10] was one of those who originated the development of the semi-empirical formulas for heat transfer at nucleate boiling (see [21]), which then have been brought (and remain there at the time being) into the standard handbooks (see, for example, [22]).

7.3.2 Periodic Model of Nucleate Boiling

Oscillations of the thickness of a liquid film. One of the possible models of nucleate boiling is considered below. This model has incorporated a minimal number of numerical constants. The basic emphasis is done on an independent validation of the separate components in the model, which in a narrow sense can be understood as a verification of the values of these constants. In order to undertake a more detailed analysis of the thermal effect of a body on heat transfer at nucleate boiling, it is necessary first of all to analyze spatial and temporal periodicity of the process. In accordance with the quasistationary character of the process of boiling, it is natural to believe that the thickness of the film will undergo periodic oscillations in time with a certain period τ_0. On the other hand, presence of the fixed sites of boiling assumes unavoidable spatial nonuniformity (waviness) of the film with a certain lengthscale (wavelength) L. It is possible with a good degree of accuracy to reflect the mentioned spatiotemporal periodicity via setting harmonic oscillations of the film thickness under the law of a progressive wave (Fig. 7.7)

$$\delta_{\mathrm{f}} = \langle \delta_{\mathrm{f}} \rangle \left\{ 1 + b \cos \left[2\pi \left(\frac{Z}{L} - \frac{\tau}{\tau_0} \right) \right] \right\}. \tag{7.11}$$

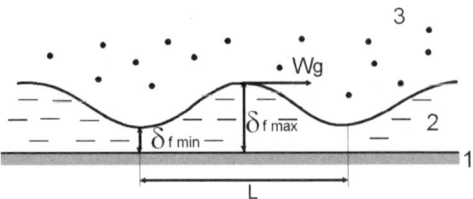

Fig. 7.7. Periodic model of nucleate boiling: (1) heated surface, (2) oscillating liquid film, (3) vapor conglomerates

As the results of visual investigations [23] show, nucleate boiling is characterized by some microroughness with a linear lengthscale of the order of magnitude comparable with the diameter of a critical vapor nucleus, which represents a lengthscale of some bubble microroughness on the heated surface. Based on this fact, let us assume that the minimal film thickness for the period of oscillations

$$\delta_{f\,min} = \langle \delta_f \rangle (1 - b) \tag{7.12}$$

becomes equal to

$$\delta_{f\,min} = 2R_*. \tag{7.13}$$

Based on the analogy with a near-wall turbulent flow [12], one can assume that the maximal film thickness

$$\delta_{f\,max} = \langle \delta_f \rangle (1 + b) \tag{7.14}$$

is proportional to the thickness of the viscous sublayer based on the nucleation rate

$$\delta_{f\,max} = \beta_1 \frac{\nu_f}{w_g}. \tag{7.15}$$

According to (7.6) and (7.11), the ATHTC can be determined in this case from the following relation

$$\langle h \rangle = \frac{k_f}{\sqrt{\delta_{f\,min} \delta_{f\,max}}}. \tag{7.16}$$

Finally, the law of heat transfer at nucleate boiling can be derived from (7.11)–(7.16)

$$q = \frac{1}{4\beta_1} \frac{k_f^2 \vartheta^3}{\nu_f \sigma T_s}. \tag{7.17}$$

For the value of the numerical constant $\beta_1 = 250$, (7.17) coincides with (7.7).
Nucleation site density. Let us consider now in more details the issue connected with the determination of the nucleation site density

$$n_F = \frac{1}{L^2}. \tag{7.18}$$

7.3 Nucleate Boiling

In the theory Labuntsov [11], this value is to be determined from (7.2) and (7.18) as

$$n_F \approx \frac{10^{-8}}{R_*^2}. \tag{7.19}$$

Equation (7.19) predicts square-law dependence of the nucleation site density on the temperature difference

$$n_F \sim \vartheta^2. \tag{7.20}$$

However, an analysis of the experimental investigations into the near-wall structure of nucleate boiling carried out up to the present time reveals that the power exponent should be given much larger numerical values [14, 15, 23, 24]

$$n_F \sim \vartheta^{3...5}. \tag{7.21}$$

The estimate (7.20) was obtained by Labuntsov grounding on an assumption that the radius of a vapor nucleus (microlengthscale) determined by (7.3) is the unique characteristic lengthscale of the entire process. Therefore, the experimental proof of the higher power exponent in this dependence indirectly points out at the existence of the second (macro)lengthscale. Another indirect evidence of the insufficiency of the Labuntsov's model consists also in the abnormally small value of the numerical constant in (7.19). A simple physical model of a flow in a near-wall liquid film on a heated surface between the boiling nucleation sites proposed below allows determining the aforementioned macrolengthscale.

Let us consider a stationary flow over the length a liquid film that directly adjoins to its thinnest part (Fig. 7.8).[1] Considering the thickness of the layer constant and equal to $\delta_{f\,min}$, one can thus receive a case of a viscous flow of a liquid in a layer with a constant suction rate on its top boundary equal to the rate of evaporation of the liquid

$$w_f = \frac{k_f \vartheta}{\delta_{f\,min} h_{fg} \rho_f}. \tag{7.22}$$

The pressure gradient in the liquid in the Z-direction can be expressed as [25]

$$\frac{dp}{dZ} = 3\frac{\mu_f w_f}{\delta_{f\,min}^3} Z. \tag{7.23}$$

As shown in the work [26] with the reference to a problem of film condensation, the flow of a liquid against viscose forces for the considered case of a very

---

[1] Apparently, for the first time the specified problem with the reference to a problem of liquid film evaporation was theoretically and experimentally investigated by the authors of the work [27]. Later the model of an evaporating liquid film was used by Straub at a research of a problem of vapor bubble dynamics on a solid wall at boiling of a liquid (see survey work [28]).

small (microscopic) film thickness can effectively exist mainly at the expense of capillary forces

$$\frac{dp}{dZ} = -\sigma \frac{dK_f}{dZ}. \quad (7.24)$$

Here K_f is curvature of the film surface, which for small values of the derivative $d\delta_f/dZ \ll 1$ can be approximately determined as

$$K_f \approx \frac{d^2\delta_f}{dZ^2}. \quad (7.25)$$

From (7.22)–(7.25), one can derive the following differential equation

$$\frac{d^3\delta_f}{dZ^3} = -AZ, \quad (7.26)$$

where

$$A = 3\frac{\nu_f k_f \vartheta}{h_{fg} \sigma \delta_{f\,min}^4}. \quad (7.27)$$

From the conditions of conjugation of a film with a layer of the bubble microroughness, two boundary conditions for (7.26) physically follow

$$Z = 0 : \delta_f = \delta_{f\,min}, \quad \frac{d\delta_f}{dZ} = 0. \quad (7.28)$$

Then, a simple integration of (7.26) leads to the following equation for the dependence of the liquid film thickness on the longitudinal coordinate

$$\delta_f = \delta_{f\,min} + \frac{CZ^2}{2} - \frac{AZ^4}{24}. \quad (7.29)$$

It follows from (7.29) that the dependence $\delta_f(Z)$ exhibits consecutively a growing branch $(d\delta_f/dZ > 0)$, an inflection point $(d^2\delta_f/dZ^2 = 0)$, a point of maximum $(d\delta_f/dZ = 0)$, and a descending branch $(d\delta_f/dZ < 0)$ (Fig. 7.8). Since the descending branch is physically unjustifiable, it is necessary to trim the dependence $\delta_f(Z)$ at a certain point, i.e., to determine in doing so both the constant C, and the effective length of the film L (or, in other words, the spatial lengthscale of periodicity $Z_0 \equiv L$). From the reasons of symmetry of the film profile (or, in other words, smooth interface between two adjacent boiling nucleation sites), let us accept that the condition of trimming is fulfilled at the point of the maximum of the dependence $\delta_f(Z)$

$$Z = \frac{L}{2} : \frac{d\delta_f}{dZ} = 0. \quad (7.30)$$

Thus, we have obtained a picture of the stopped progressive wave

$$\delta_f = \langle \delta_f \rangle \left[1 + b\cos\left(2\pi\frac{Z}{L}\right)\right]. \quad (7.31)$$

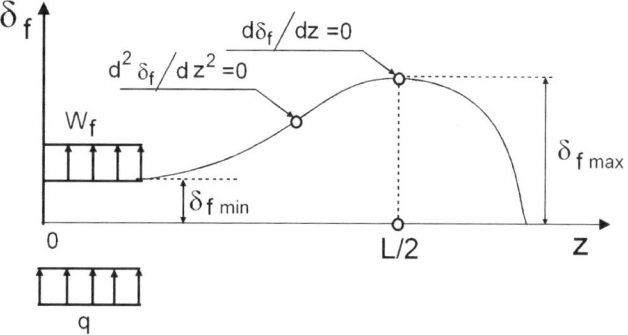

Fig. 7.8. Illustration to the determination of the effective length of the liquid film

One can further find out from (7.28)–(7.30) that

$$\frac{L}{\delta_{f\,min}} = \beta_2 \left(\frac{h_{fg}\sigma\delta_{f\,max}}{\nu_f k_f \vartheta}\right)^{1/4}. \tag{7.32}$$

One should point out that the estimate of the relation between the maximal and minimal thickness of the evaporating liquid film ($\delta_{f\,min} \ll \delta_{f\,max}$) suggested in the works [27, 28] was used at the derivation of (7.32). An interrelation between the macro-L and micro-$\delta_{f\,min}$ lengthscales of the process of nucleate boiling can be deduced from (7.15) and (7.32)

$$\frac{L}{\delta_{f\,min}} = \beta_3 \frac{(\nu_f \rho_g T_s)^{1/4} (h_{fg}\sigma)^{1/2}}{k_f^{3/4}\vartheta}. \tag{7.33}$$

For a transition from the frozen [(7.31)] to the running [(7.11)] progressive wave of oscillations of the film thickness, it is necessary to find out the period of temporal oscillations. It is natural to believe that oscillations of the heat transfer intensity extend along the surface of a body with a phase speed of the order of magnitude comparable with the vapor nucleation rate [(7.10)]. Then the timescale of periodicity can be determined from the relation

$$\tau_0 = \frac{L}{w_g}. \tag{7.34}$$

Knowing macroscale L, it is possible to determine the required nucleation site density

$$n_F = \beta_4 \frac{h_{fg}(\lambda_f \rho_g)^{3/2} \vartheta^4}{\nu_f^{1/2} T_s^{5/2} \sigma^3}. \tag{7.35}$$

As one can conclude from (7.35), the theoretical model represented here provides a qualitatively true dependence of the nucleation site density on the temperature difference and agrees well with the correspondent tendencies documented in works [14, 15, 23, 24].

Factor of conjugation. Having a clear hydrodynamic picture of nucleate boiling, let us pass now to the determination of the FC. Let us consider for clarity a case of a semi-infinite body with the wall thickness much larger than the wavelength L of oscillations. Then, for oscillations of the THTC in accordance with the inverse harmonic law analyzed here, the FC can be determined by (4.22), which can be rewritten in the following form

$$\varepsilon = \frac{1}{\sqrt{1 + 2B/\varepsilon_{\min} + B^2} - B}. \tag{7.36}$$

Equation (7.36) includes the following dimensionless parameters:

- $B = \frac{\langle \bar{h} \rangle}{(1+m^2)^{1/4}}$,
- $\langle \bar{h} \rangle = \frac{\langle h \rangle L}{k}$, the Biot number
- $m = \frac{L^2}{\alpha \tau_0}$, inverse Fourier number
- $\varepsilon_{\min} = 2\frac{\sqrt{\kappa}}{1+\kappa}$, the minimal value of the factor of conjugation
- $\kappa = \frac{\delta_{f\,\min}}{\delta_{f\,\max}}$, the ratio of the minimal and maximal thickness of the film

The results for the characteristics of the flow of a liquid film between the boiling nucleation sites obtained above allow writing down the specified parameters in the following form:

- $\langle \bar{h} \rangle = \beta_5 \frac{k_f}{k} \Psi$, the Biot number
- $m = \beta_6 \frac{\alpha_f}{\alpha} \Psi \, Pr \, J$, inverse Fourier number
- $\kappa = \beta_7 J^2$, the ratio of the minimal and maximal thickness of the film

In turn, three new (primary) parameters appear here:

- $Pr = \frac{\nu_f}{\alpha_f}$, the Prandtl number for a liquid
- $J = \frac{k_f \vartheta}{\nu_f h_{fg} \rho_g}$, modified Jakob number
- $\Psi = \frac{(k_f T_s)^{1/4}}{(\nu_f \rho_g)^{3/4}} \left(\frac{\sigma}{h_{fg}} \right)^{1/2}$, dimensionless complex.

Thus, the phenomenon of conjugate convective–conductive heat transfer at nucleate boiling represents an essentially multiple-parameter problem. Depending on a combination of the determining parameters, it comprises a wide spectrum of subproblems, which can noticeably differ from each other in the quantitative and qualitative aspects.

Influence of thermophysical properties of a body. If thermal conductivity or thermal diffusivity of a body tends to infinity, one has a limiting case of absence of the conjugation:

$$\left. \begin{array}{c} \frac{k}{k_f} \to \infty \\ \frac{\alpha}{\alpha_f} \to \infty \end{array} \right\} \Rightarrow \varepsilon \to 1. \tag{7.37}$$

In turn, for a case of negligibly small values of the specified parameters, the thermal effect of a body reaches its maximum:

$$\left.\begin{array}{r}\frac{k}{k_f} \to 0 \\ \frac{\alpha}{\alpha_f} \to 0\end{array}\right\} \Rightarrow \varepsilon \to \varepsilon_{\min}. \tag{7.38}$$

As it was already repeatedly pointed out above, such character of the behavior of the FC is believed to be natural and remains in the mainstream of the theory developed in the present book.

Influence of pressure. As the pressure decreases, the modified Jacob number increases. This leads to the growth of the parameter κ, i.e., to the diminishing the distinctions between the minimal and maximal thickness of a film. Physically it corresponds to the decrease in the amplitude of oscillations of the THTC, i.e., to weakening of the thermal effect of the wall. It is interesting to point out that the present theory predicts full disappearance of the influence of a body on heat transfer at some limiting value of the modified Jacob number equal to

$$J_{\max} = \beta_7^{-1/2} \Rightarrow \kappa = \varepsilon_{\min} = \varepsilon = 1. \tag{7.39}$$

This fact obviously agrees with the tendency, which has been revealed in the experiments [29–31] on boiling of nitrogen [29,30], water and ethanol [31]. Thus, in the present research for the first time (as far as it is known to the author) a physical explanation is given for the tendency of the degeneration of the wall's thermal effect on heat transfer at nucleate boiling at the decreasing pressure.

Numerical constants. A prominent feature of the theoretical model presented above is a unique and rare opportunity of the exact determination of all the numerical constants involved in the model:

- $\beta_1 = 250$
- $\beta_2 = 3.364$
- $\beta_3 = 7.521$
- $\beta_4 = 1.105 \times 10^{-5}$
- $\beta_5 = \beta_6 = 4.788 \times 10^{-2}$
- $\beta_7 = 1.6 \times 10^{-5}$

Summary. Certainly, the conjugate convective–conductive problem for nucleate boiling cannot be in principle comprehensively solved within the framework of the simplified scheme presented above. The author realizes in full that carrying out of a separate extensive research is necessary in order to undertake any kind of a representative analysis of this highly complicated physical process. The basic fundamentals of the theoretical model of nucleate boiling of a liquid under conditions of natural convection outlined in the present chapter have been published by the author in the works [32–43]

References

1. Mikic BB (1969) On mechanism on dropwise condensation. Int J Heat Mass Transf 12: 1311–1323.
2. Griffith P, Lee MS (1967) The effect of surface thermal properties and finish on dropwise condensation. Int J Heat Mass Transf 10: 697–707.
3. Wilkins D, Bromley L (1973) Dropwise condensation phenomena. AIChE J 19: 839–845.
4. Hannemann RJ, Mikic BB (1976) An analysis of the effect of surface thermal conductivity on the rate of heat transfer in dropwise condensation. Int J Heat Mass Transf 19: 1299–1307.
5. Hannemann RJ, Mikic BB (1976) An experimental investigation into the effect of surface thermal conductivity on the rate of heat transfer in dropwise condensation. Int J Heat Mass Transf 19: 1309–1317.
6. Hannemann RJ (1978) Condensing surface thickness effects in dropwise condensation. Int J Heat Mass Transf 21: 65–66.
7. Rose JW (1967) Further aspects of dropwise condensation theory. Int J Heat Mass Transf 10: 697–707.
8. Rose JW (2002) Dropwise condensation theory and experiment: a review. Proc Inst Mech Eng A J Power Energy 2: 115–128.
9. Rose JW (2003) Heat-transfer coefficients, Wilson plots and accuracy of thermal measurements. Exp Therm Fluid Sci 28: 3–12.
10. Stephan K (1992) Heat Transfer in Condensation and Boiling. Springer, Berlin Heidelberg New York.
11. Labuntsov DA (2000) Physical Principles of Energetics. Selected Papers. Power Engineering Institute, Moscow (in Russian).
12. Schlichting H, Gersten K (1997) Grenzschicht-Theorie. Springer, Berlin Heidelberg New York.
13. Pioro IL, Rohsenow W, Doerffer SS (2004) Nucleate pool-boiling heat transfer. I. Review of parametric effects of boiling surface. Int J Heat Mass Transf 47: 5033–5044.
14. Kenning D, Golobiq I, Xing H, Baselj M, Lojk V, von Hardenberg J (2006) Mechanistic models for pool nucleate boiling heat transfer: input and validation. Heat Mass Transf 42: 511–527.
15. Dhir VK (2006) Mechanistic prediction of nucleate boiling heat transfer–achievable or a hopeless task? ASME J Heat Transfer 123: 1–12.
16. Dhir VK (2001) Numerical simulations of pool-boiling heat transfer. AIChE J 47: 813–834.
17. Yu B, Cheng P (2002) A fractal model for nucleate pool boiling heat transfer. ASME J Heat Transf 124: 1117–1124.
18. Mandelbrot BB (1982) The Fractal Geometry of Nature. Freeman WH (ed.). New York.
19. Application of Fractals and Chaos (1993). Eanshaw RA (ed.). Springer, Berlin.
20. Cebeci T Turbulence Models and Their Application (2003). Springer, Berlin Heidelberg New York.
21. Stephan K (1963) Mechanismus und Modellgesetz des Wärmeübergangs bei der Blasenverdampfung. Chemie-Ingenieur-Technik 35 (11) : 775–784.
22. Gorenflo D (2002) Behältersieden (Sieden bei freier Konvektion). VDI – Wärmeatlas, Hab. Springer, Berlin Heidelberg New York.

23. Yusen Qi Y, Klausner (2006) Comparison of nucleation site density for pool boiling and gas nucleation. ASME J Heat Transf 128: 13–20.
24. Benjamin RJ, Balakrishnan AR (1997) Nucleation site density in pool boiling of saturated pure liquids: effect of surface microroughness and surface and liquid physical properties. Exp Thermal Fluid Sci 15: 32–42.
25. Sherman FS (1990) Viscous Flow, McGraw-Hill.
26. Rose JW (2004) Surface tension effects and enhancement of condensation heat transfer. Trans IChemE Part A Chem Eng Res Design 82: 419–429.
27. Wayner PC, Kao YK, LaCroix LV (1976) The interline heat transfer coefficient on an evaporating wetting film. Int J Heat Mass Transf 19: 487–492.
28. Straub J. (2001) Boiling heat transfer and bubble dynamics in microgravity. Adv Heat Transf 35: 57–172.
29. Dudkevich AC, Akhmedov FD (1974) Experimental study of influence of thermophysical properties of heating surface on boiling of nitrogen at elevated pressures. Works of Moscow Power Engineering Institute Issue 198: 41–47. (in Russian).
30. Kirichenko YA, Rusanov KV, Tyurina EG (1985) Effect of pressure on heat exchange in nitrogen boiling under conditions of free motion in an annular channel. J Eng Phys Thermophys 49: 1005–1010.
31. Gorodov AK, Kaban'kov ON, Martynov YK, Yagov VV (1979) Effect of material and of the thickness of the heating surface on the heat transfer rate in boiling of water and ethanol at subatmospheric pressures. J Heat Transf Soviet Res 11(3): 44–52.
32. Zudin YB (1992) Analog of the Rayleigh equation for the problem of bubble dynamics in a tube. J Eng Phys Thermophys 63: 672–675.
33. Zudin YB (1993) The calculation of parameters of the evaporating meniscus a thin liquid film. High Temp 31: 714–716.
34. Zudin YB (1997) The use of the model of evaporating macrolayer for determining the characteristics of nucleate boiling. High Temp 35: 565–571.
35. Zudin YB (1997) Calculation of critical thermal loads under extreme intensities of mass forces. Heat Transf Res 28: 481–483.
36. Zudin YB (1997) Influence of the coefficient of thermal activity of a wall on heat transfer in transient boiling. J Eng Phys Thermophys 71: 696–698.
37. Zudin YB (1997) Law of vapor-bubble growth in a tube in the region of low pressures. J Eng Phys Thermophys 70: 714–717.
38. Zudin YB (1998) The distance between nucleate boiling sites. High Temp 36: 662–663.
39. Zudin YB (1998) Calculation of the surface density of nucleation sites in nucleate boiling of a liquid. J Eng Phys Thermophys 71: 178–183.
40. Zudin YB (1998) Boiling of liquid in the cell of a jet printer. J Eng Phys Thermophys 71: 217–220.
41. Zudin YB (1999) Burn-out of a liquid under conditions of natural convection. J Eng Phys Thermophys 72: 50–53.
42. Zudin YB (1999) Wall non-isothermicity effect on the heat exchange in jet reflux. J Eng Phys Thermophys 72: 309–312.
43. Zudin YB (1999) Model of heat transfer in bubble boiling. J Eng Phys Thermophys 72: 438–444.

A

Proof of the Fundamental Inequalities

A.1 Proof of the First Fundamental Inequality

It is required to prove the inequality:

$$\varepsilon \leq 1. \tag{A.1}$$

By definition, we have at our disposal the following relations:

(a) For the THTC

$$h = \frac{q_\delta}{\vartheta_\delta} = \frac{\langle q_\delta \rangle + \hat{q}_\delta}{\langle \vartheta_\delta \rangle + \hat{\vartheta}_\delta}, \tag{A.2}$$

(b) For the EHTC

$$h_\mathrm{m} = \frac{\langle q_\delta \rangle}{\langle \vartheta_\delta \rangle}. \tag{A.3}$$

Here $\vartheta_\delta, q_\delta$ are the local values of the temperature and heat flux at $X = \delta$, $\langle \vartheta_\delta \rangle, \langle q_\delta \rangle$ are their average values, $\hat{\vartheta}_\delta, \hat{q}_\delta$ are their oscillating values.
Introducing the correspondent normalized values

$$\tilde{\vartheta} = \frac{\hat{\vartheta}_\delta}{\langle \vartheta_\delta \rangle}, \quad \tilde{q} = \frac{\hat{q}_\delta}{\langle q_\delta \rangle}, \tag{A.4}$$

let us write down (A.2) as

$$h = h_\mathrm{m} \frac{1 + \tilde{q}_\delta}{1 + \tilde{\vartheta}_\delta}, \tag{A.5}$$

then further rewrite as the expression

$$\varepsilon = \frac{(1 + \psi)\left(1 + \tilde{\vartheta}_\delta\right)}{1 + \tilde{q}_\delta}, \tag{A.6}$$

A Proof of the Fundamental Inequalities

and, at last, reduce it to the form equivalent to the relation (2.29)

$$\varepsilon (1 + \tilde{q}_\delta) = (1 + \psi) \left(1 + \tilde{\vartheta}_\delta\right). \tag{A.7}$$

Averaging of (A.7) over the period of the variation of the progressive wave gives

$$\varepsilon = 1 + \left\langle \psi \tilde{\vartheta}_\delta \right\rangle. \tag{A.8}$$

Multiplying both parts of (A.7) by the value $\left(1 + \tilde{\vartheta}_\delta\right)$ and averaging the result, one can obtain

$$\varepsilon \left(1 + \left\langle \tilde{\vartheta}_\delta \tilde{q}_\delta \right\rangle \right) = 1 + 2\left\langle \tilde{\vartheta}_\delta \psi \right\rangle + \left\langle \tilde{\vartheta}_\delta^2 (1 + \psi) \right\rangle, \tag{A.9}$$

or, with allowance for (A.6),

$$\varepsilon = \frac{1 - \left\langle \tilde{\vartheta}_\delta^2 (1 + \psi) \right\rangle}{1 - \left\langle \tilde{\vartheta}_\delta \tilde{q}_\delta \right\rangle}. \tag{A.10}$$

As the obvious inequalities

$$(1 + \psi) \geq 0, \quad \tilde{\vartheta}_\delta^2 \geq 0, \tag{A.11}$$

always hold, then, hence, the following inequality should be also valid

$$\left\langle \tilde{\vartheta}_\delta^2 (1 + \psi) \right\rangle \geq 0. \tag{A.12}$$

It follows from here that for the first basic inequality (A.1) to hold, it is sufficient to provide validity of the following inequality

$$\left\langle \tilde{\vartheta}_\delta \tilde{q}_\delta \right\rangle \leq 0. \tag{A.13}$$

Let us write down the heat conduction equation for a plate

$$c\rho \frac{\partial \hat{\vartheta}}{\partial \tau} = k \left(\frac{\partial^2 \hat{\vartheta}}{\partial X^2} + \frac{\partial^2 \hat{\vartheta}}{\partial Z^2} \right), \tag{A.14}$$

which, in view of the law of Fourier, can be rewritten in the following form

$$c\rho \frac{\partial \hat{\vartheta}}{\partial \tau} = - \left(\frac{\partial \hat{q}_x}{\partial X} + \frac{\partial \hat{q}_z}{\partial Z} \right) \tag{A.15}$$

that, with allowance for (A.3), can be transformed as:

$$c\rho \frac{\partial \tilde{\vartheta}}{\partial \tau} = -h_m \left(\frac{\partial \tilde{q}_x}{\partial X} + \frac{\partial \tilde{q}_z}{\partial Z} \right). \tag{A.16}$$

A.1 Proof of the First Fundamental Inequality

Multiplying both parts of (A.16) by the value $\tilde{\vartheta}$, one can obtain:

$$c\rho \frac{\partial \left(\tilde{\vartheta}^2/2\right)}{\partial \tau} + h_m \tilde{\vartheta}\left(\frac{\partial \tilde{q}_x}{\partial X} + \frac{\partial \tilde{q}_z}{\partial Z}\right) = 0. \quad \text{(A.17)}$$

Let us write down an identity:

$$\tilde{\vartheta}\frac{\partial \tilde{q}_x}{\partial X} \equiv \frac{\partial \left(\tilde{\vartheta}\tilde{q}_x\right)}{\partial X} - \tilde{q}_x \frac{\partial \tilde{\vartheta}}{\partial X}, \quad \text{(A.18)}$$

which, using the law of Fourier, can be rewritten in the following form

$$\frac{\partial \tilde{\vartheta}}{\partial X} \equiv -\frac{h_m}{k}\tilde{q}_x, \quad \text{(A.19)}$$

$$\tilde{\vartheta}\frac{\partial \tilde{q}_x}{\partial X} = \frac{\partial \left(\tilde{\vartheta}\tilde{q}_x\right)}{\partial X} + \frac{h_m}{k}\tilde{q}_x^2. \quad \text{(A.20)}$$

Performing the same procedure with the term $\tilde{\vartheta}\left(\partial \tilde{q}_z/\partial Z\right)$, we shall further rewrite (A.17) as:

$$\frac{c\rho}{h_m}\frac{\partial \left(\tilde{\vartheta}^2/2\right)}{\partial \tau} + \frac{\partial \left(\tilde{q}_x\tilde{\vartheta}\right)}{\partial X} + \frac{\partial \left(\tilde{q}_z\tilde{\vartheta}\right)}{\partial Z} + \frac{h_m}{k}\left(\tilde{q}_x^2 + \tilde{q}_z^2\right) = 0. \quad \text{(A.21)}$$

Let us integrate the left-hand side of (A.21) over X within limits from 0 to δ:

$$\frac{1}{2}\frac{c\rho}{h_m}\frac{\partial}{\partial \tau}\int_0^\delta \tilde{\vartheta}^2 \mathrm{d}X + \left(\tilde{q}_x\tilde{\vartheta}\right)\Big|_0^\delta + \frac{\partial}{\partial Z}\int_0^\delta \left(\tilde{q}_z\tilde{\vartheta}\right)\mathrm{d}X$$

$$+ \frac{h_m}{k}\int_0^\delta \left(\tilde{q}_x^2 + \tilde{q}_z^2\right)\mathrm{d}X = 0. \quad \text{(A.22)}$$

Let us express the required value $\tilde{q}_x\tilde{\vartheta}_\delta \equiv \tilde{\vartheta}_\delta \tilde{q}_\delta$ from (A.22)

$$\tilde{\vartheta}_\delta \tilde{q}_\delta = \tilde{\vartheta}_0 \tilde{q}_0 - \frac{1}{2}\frac{c\rho}{h_m}\frac{\partial}{\partial \tau}\int_0^\delta \tilde{\vartheta}^2 \mathrm{d}X - \frac{\partial}{\partial Z}\int_0^\delta \left(\tilde{q}_z\tilde{\vartheta}\right)\mathrm{d}X$$

$$- \frac{h_m}{k}\int_0^\delta \left(\tilde{q}_x^2 + \tilde{q}_z^2\right)\mathrm{d}X. \quad \text{(A.23)}$$

Averaging both parts of (A.23) over the spatial coordinate Z and time τ (that is equivalent to averaging over the period of variation of the progressive wave), one can notice that in doing so the second and third terms in the right-hand side of this equation drop out. From here, one can obtain the following equation

$$\left\langle \tilde{\vartheta}_\delta \tilde{q}_\delta \right\rangle = \left\langle \tilde{\vartheta}_0 \tilde{q}_0 \right\rangle - \frac{h_m}{k} \int_0^\delta \left\langle \tilde{q}_x^2 + \tilde{q}_z^2 \right\rangle \mathrm{d}X. \tag{A.24}$$

The further steps of the mathematical proof will be carried out separately for each of the respective TBC.

TBC: $\vartheta_0 = \mathrm{const}, q_0 = \mathrm{const}$. In this case, one can have either $\tilde{\vartheta}_0 = 0$, or $\tilde{q}_0 = 0$. Hence, it can be concluded that $\left\langle \tilde{\vartheta}_0 \tilde{q}_0 \right\rangle = 0$, that results in the following equation

$$\left\langle \tilde{\vartheta}_\delta \tilde{q}_\delta \right\rangle = -\frac{h_m}{k} \int_0^\delta \left\langle \tilde{q}_x^2 + \tilde{q}_z^2 \right\rangle \mathrm{d}X. \tag{A.25}$$

Since the following inequality is obviously valid

$$\left\langle \tilde{q}_x^2 + \tilde{q}_z^2 \right\rangle \geq 0, \tag{A.26}$$

then inequality (A.13) follows inevitably from (A.25), and this is actually what had to be proved.

TBC $h_0 = \mathrm{const}$. In this case, $\hat{q}_0 = -h_0 \hat{\vartheta}_0$ or

$$\tilde{q}_0 = -\frac{h_0}{h_m} \tilde{\vartheta}_0. \tag{A.27}$$

Multiplying both parts of (A.27) by value $\tilde{\vartheta}_0$ and averaging the resulting expression, one can obtain

$$\left\langle \tilde{\vartheta}_0 \tilde{q}_0 \right\rangle = -\frac{h_0}{h_m} \left\langle \tilde{\vartheta}_0^2 \right\rangle. \tag{A.28}$$

As the obvious inequality is valid

$$\left\langle \tilde{\vartheta}_0^2 \right\rangle \geq 0, \tag{A.29}$$

then from (A.27) the inequality follows

$$\left\langle \tilde{\vartheta}_0 \tilde{q}_0 \right\rangle \leq 0. \tag{A.30}$$

Having substituted the value $\left\langle \tilde{\vartheta}_0 \tilde{q}_0 \right\rangle$ from (A.29) into (A.12), one can obtain

$$\left\langle \tilde{\vartheta}_\delta \tilde{q}_\delta \right\rangle = -\frac{h_0}{h_m} \left\langle \tilde{\vartheta}_0^2 \right\rangle - \frac{h_m}{k} \int_0^\delta \left\langle \tilde{q}_x^2 + \tilde{q}_z^2 \right\rangle \mathrm{d}X. \tag{A.31}$$

A.1 Proof of the First Fundamental Inequality

From here, the inequality (A.13) results, and this is in fact what had to be proved.

TBC: contact to the second wall. Making with the heat conduction equation for the second plate the same transformations, as those done for the first plate, one can derive the following equation:

$$\left\langle \tilde{\vartheta}_0 \tilde{q}_0 \right\rangle = \left\langle \tilde{\vartheta}_1 \tilde{q}_1 \right\rangle - \frac{h_m}{k_1} \int_0^\delta \left\langle \tilde{q}_{x1}^2 + \tilde{q}_{z1}^2 \right\rangle \mathrm{d}X_1. \qquad (A.32)$$

Here the subscript "1" relates to the second plate, $\tilde{\vartheta}_1, \tilde{q}_1$ designate oscillating components of the temperatures and heat fluxes on the external surface of the second plate. Let us rewrite (A.24) with allowance for (A.32) in the following form

$$\left\langle \tilde{\vartheta}_\delta \tilde{q}_\delta \right\rangle = \left\langle \tilde{\vartheta}_1 \tilde{q}_1 \right\rangle - \frac{h_m}{k_1} \int_0^\delta \left\langle \tilde{q}_{x1}^2 + \tilde{q}_{z1}^2 \right\rangle \mathrm{d}X_1$$

$$- \frac{h_m}{k} \int_0^\delta \left\langle \tilde{q}_x^2 + \tilde{q}_z^2 \right\rangle \mathrm{d}X. \qquad (A.33)$$

On the external surface of the second plate, one should preset one of the following TBC: either $\vartheta_0 = \mathrm{const}$ or $q_0 = \mathrm{const}$ or $h_0 = \mathrm{const}$. However, for all the three mentioned kinds of the boundary condition, validity of the inequality

$$\left\langle \tilde{\vartheta}_1 \tilde{q}_1 \right\rangle \leq 0. \qquad (A.34)$$

was proved above. From here, with allowance for (A.13), the inequality (A.13) follows, and this is effectively what had to be proved.

Plate, cylinder and sphere (generalized case). It is required to prove validity of the very same inequality (A.13). Let us write down the heat conduction equation for a body of a generalized geometry (plate, cylinder, sphere) in the following form

$$c\rho \frac{\partial \hat{\vartheta}}{\partial \tau} = k \left(\frac{\partial^2 \hat{\vartheta}}{\partial X^2} + \frac{s}{X} \frac{\partial \hat{\vartheta}}{\partial X} \right). \qquad (A.35)$$

Here X is the cross-sectional coordinate counted from the plane of symmetry (for a plate), the axis of symmetry (for a cylinder) or the center of symmetry (for a sphere); s is a geometrical factor equal to: $s = 0$ for a plate, $s = 1$ for a cylinder, $s = 2$ for a sphere. Let us rewrite (A.19) with the help of the law of Fourier as:

$$c\rho \frac{\partial \hat{\vartheta}}{\partial \tau} = - \left(\frac{\partial \hat{q}}{\partial X} + \frac{s}{X} \hat{q} \right) \qquad (A.36)$$

A Proof of the Fundamental Inequalities

or as

$$c\rho \frac{\partial \tilde{\vartheta}}{\partial \tau} = -h_m \left(\frac{\partial \tilde{q}}{\partial X} + \frac{s}{X} \tilde{q} \right). \tag{A.37}$$

Multiplying both parts of (A.37) by the value $\tilde{\vartheta}$, one can further derive:

$$\frac{1}{2} c\rho \frac{\partial \left(\tilde{\vartheta}^2 \right)}{\partial \tau} + h_m \tilde{\vartheta} \left(\frac{\partial \tilde{q}}{\partial X} + \frac{s}{X} \tilde{q} \right) = 0. \tag{A.38}$$

Let us write down an identity:

$$\tilde{\vartheta} \frac{\partial \tilde{q}}{\partial X} \equiv \frac{\partial \left(\tilde{\vartheta} \tilde{q} \right)}{\partial X} - \tilde{q} \frac{\partial \tilde{\vartheta}}{\partial X}, \tag{A.39}$$

or, using the law of Fourier

$$\frac{\partial \tilde{\vartheta}}{\partial X} \equiv -\frac{h_m}{k} \tilde{q}, \tag{A.40}$$

$$\tilde{\vartheta} \frac{\partial \tilde{q}}{\partial X} = \frac{\partial \left(\tilde{\vartheta} \tilde{q} \right)}{\partial X} + \frac{h_m}{k} \tilde{q}^2. \tag{A.41}$$

Let us then reduce the equation of heat conduction (A.38) with the allowance for (A.41) to the following form:

$$\frac{1}{2} \frac{c\rho}{h_m} \frac{\partial \left(\tilde{\vartheta}^2 \right)}{\partial \tau} + \frac{1}{X^s} \frac{\partial \left(X^s \tilde{q} \tilde{\vartheta} \right)}{\partial X} + \frac{h_m}{k} \tilde{q}^2 = 0. \tag{A.42}$$

Let us further multiply the left-hand side of (A.42) by the value X^r and integrate the resulting expression over X within the limits from 0 up to δ:

$$\frac{1}{2} \frac{c\rho}{h_m} \frac{\partial}{\partial \tau} \int_0^\delta \tilde{\vartheta}^2 X^s dX + \left(\tilde{q} \tilde{\vartheta} X^s \right)\Big|_0^\delta + \frac{h_m}{k} \int_0^\delta \tilde{q}^2 X^s dX = 0. \tag{A.43}$$

From (A.43), one can derive the following equation:

$$\tilde{\vartheta}_\delta \tilde{q}_\delta \delta^s = -\frac{1}{2} \frac{c\rho}{h_m} \frac{\partial}{\partial \tau} \int_0^\delta \tilde{\vartheta}^2 X^s dX - \frac{h_m}{k} \int_0^\delta \tilde{q}^2 X^s dX. \tag{A.44}$$

In doing so, the trivial equality $\tilde{q}_0 = 0$ was taken into account, which follows from the condition of symmetry of the oscillating temperature field. Averaging both parts of (A.44) over the variable τ and noticing that the first term in the right-hand side drops out as a result of this procedure, one can obtain:

$$\left\langle \tilde{\vartheta}_\delta \tilde{q}_\delta \right\rangle = -\frac{h_m}{k \delta^s} \int_0^\delta \tilde{q}^2 X^s dX. \tag{A.45}$$

From the last equation, the inequality (A.13) finally follows, and this is actually what had to be proved.

A.2 Proof of the Second Fundamental Inequality

It is required to prove a fundamental inequality:

$$\varepsilon \geq \left\langle \frac{1}{1+\psi} \right\rangle^{-1}. \tag{A.46}$$

Having divided both parts of (A.7) by the value $\varepsilon(1+\psi)$, one can obtain the expression

$$\varepsilon^{-1}\left(1+\tilde{\vartheta}_\delta\right) = \frac{1+\tilde{q}_\delta}{1+\psi}. \tag{A.47}$$

Averaging of (A.47) gives:

$$\varepsilon^{-1} = \left\langle \frac{1}{1+\psi} \right\rangle + \left\langle \frac{\tilde{q}_\delta}{1+\psi} \right\rangle. \tag{A.48}$$

Having divided both parts of (A.48) by the value $\langle 1/(1+\psi)\rangle$, one can further obtain

$$\varepsilon^{-1}\left\langle \frac{1}{1+\psi} \right\rangle^{-1} = 1 + \left\langle \frac{1}{1+\psi} \right\rangle^{-1}\left\langle \frac{\tilde{q}_\delta}{1+\psi} \right\rangle. \tag{A.49}$$

It is required to prove validity of the inequality (A.46) or of the following inequality equivalent to (A.46)

$$\varepsilon^{-1}\left\langle \frac{1}{1+\psi} \right\rangle^{-1} \leq 1. \tag{A.50}$$

It follows from (A.49) that a condition sufficient to provide validity of the inequality (A.50) can be expressed as

$$\left\langle \frac{1}{1+\psi} \right\rangle^{-1}\left\langle \frac{\tilde{q}_\delta}{1+\psi} \right\rangle \leq 0. \tag{A.51}$$

Multiplying both parts of (A.47) by the value $(1+\tilde{q}_\delta)$ and averaging these both parts over time, one can obtain

$$\varepsilon^{-1}\left(1+\langle\tilde{\vartheta}_\delta\tilde{q}_\delta\rangle\right) = \left\langle \frac{1}{1+\psi} \right\rangle + 2\left\langle \frac{\tilde{q}_\delta}{1+\psi} \right\rangle + \left\langle \frac{\tilde{q}_\delta^2}{1+\psi} \right\rangle. \tag{A.52}$$

The (A.52), with the allowance for (A.48), can be written down as:

$$\varepsilon^{-1} = \frac{\langle 1/(1+\psi)\rangle - \langle \tilde{q}_\delta^2/(1+\psi)\rangle}{1 - \langle\tilde{\vartheta}_\delta\tilde{q}_\delta\rangle}. \tag{A.53}$$

146 A Proof of the Fundamental Inequalities

Having divided both parts of (A.71) by the value $\langle 1/(1+\psi)\rangle$, one can derive the following expression:

$$\varepsilon^{-1}\left\langle \frac{1}{1+\psi}\right\rangle^{-1} = \frac{1 - \langle 1/(1+\psi)\rangle^{-1}\left\langle \tilde{q}_\delta^2/(1+\psi)\right\rangle}{1 - \left\langle \tilde{\vartheta}_\delta \tilde{q}_\delta\right\rangle}. \qquad (A.54)$$

It follows from validity of the obvious inequalities

$$\left\langle \frac{1}{1+\psi}\right\rangle \geq 0, \qquad (A.55)$$

$$\left\langle \frac{\tilde{q}_\delta^2}{1+\psi}\right\rangle \geq 0 \qquad (A.56)$$

that the relation

$$\left\langle \frac{1}{1+\psi}\right\rangle^{-1}\left\langle \frac{\tilde{q}_\delta^2}{1+\psi}\right\rangle \geq 0. \qquad (A.57)$$

is also valid. With the allowance for (A.54), it follows from here that the condition sufficient for the inequality (A.50) to hold consists in the validity of the inequality (A.13). However, the validity of the inequality (A.13) for all the considered cases was proved above. Hence, as a result, validity of the second basic inequality (A.46) should be considered proved as well.

B
Functions of the Wall Thickness

Definition of functions of the wall thickness. Eigenfunctions B_n, B_n^* are introduced by the following relations

$$\left. \begin{array}{l} B_n = (r_n + is_n)\coth\left[(r_n + is_n)\bar{\delta}\right] \\ B_n^* = (r_n - is_n)\coth\left[(r_n - is_n)\bar{\delta}\right] \end{array} \right\} \Rightarrow \vartheta_0 = \text{const}, \qquad (B.1)$$

$$\left. \begin{array}{l} B_n = (r_n + is_n)\tanh\left[(r_n + is_n)\bar{\delta}\right] \\ B_n^* = (r_n - is_n)\tanh\left[(r_n - is_n)\bar{\delta}\right] \end{array} \right\} \Rightarrow q_0 = \text{const}. \qquad (B.2)$$

Here

$$r_n = \frac{n}{\sqrt{2}}\left[\sqrt{1+\left(\frac{m}{n}\right)^2}+1\right]^{1/2}, \quad s_n = \frac{n}{\sqrt{2}}\left[\sqrt{1+\left(\frac{m}{n}\right)^2}-1\right]^{1/2}, \quad m = \frac{Z_0^2}{\alpha\tau_0}.$$

Let us present the complex conjugate values B_n, B_n^* as the sums $B_n = F_n + i\Phi_n$, $B_n^* = F_n - i\Phi_n$ and further write down the correspondent functions of thickness F_n, Φ_n for two alternative TBC:

$$\left. \begin{array}{l} F_n = \dfrac{r_n \sinh(2r_n\bar{\delta}) + s_n \sin(2s_n\bar{\delta})}{\cosh(2r_n\bar{\delta}) - \cos(2s_n\bar{\delta})} \\[2mm] \Phi_n = \dfrac{s_n \sinh(2r_n\bar{\delta}) - r_n \sin(2s_n\bar{\delta})}{\cosh(2r_n\bar{\delta}) - \cos(2s_n\bar{\delta})} \end{array} \right\} \Rightarrow \vartheta_0 = \text{const}, \qquad (B.3)$$

$$\left. \begin{array}{l} F_n = \dfrac{r_n \sinh(2r_n\bar{\delta}) - s_n \sin(2s_n\bar{\delta})}{\cosh(2r_n\bar{\delta}) + \cos(2s_n\bar{\delta})} \\[2mm] \Phi_n = \dfrac{s_n \sinh(2r_n\bar{\delta}) + r_n \sin(2s_n\bar{\delta})}{\cosh(2r_n\bar{\delta}) + \cos(2s_n\bar{\delta})} \end{array} \right\} \Rightarrow q_0 = \text{const}. \qquad (B.4)$$

The functions of thickness for the limiting cases, with respect to parameter m, are considered below.

B.1 Spatial Type of Oscillations

The spatial type of oscillations of the THTC is characterized by parameters: $m = 0, \bar{\delta} = \delta/Z_0$. Let us consider the functions of thickness for the alternative TBC.

(a) TBC – $\vartheta_0 = $ const:

$$F_n = n \coth\left(n\bar{\delta}\right), \quad \Phi_n = 0. \tag{B.5}$$

An asymptotical case of a semi-infinite body

$$\bar{\delta} \to \infty : F_n \to n. \tag{B.6}$$

An asymptotical case of a negligibly thin plate

$$\bar{\delta} \to 0 : F_n \to \bar{\delta}^{-1}. \tag{B.7}$$

(b) TBC – $q_0 = $ const

$$F_n = n \tanh\left(n\bar{\delta}\right), \quad \Phi_n = 0. \tag{B.8}$$

The asymptotical case of a semi-infinite body

$$\bar{\delta} \to \infty : F_n \to n. \tag{B.9}$$

The asymptotical case of a negligibly thin plate

$$\bar{\delta} \to 0 : F_n \to n^2 \bar{\delta}. \tag{B.10}$$

B.2 Temporal Type of Oscillations

The temporal type of oscillations of the THTC is characterized by the parameters: $m = \infty, \tilde{\delta} = \delta/\sqrt{a\tau_0}$. Let us consider the functions of thickness for the alternative TBC.

(a) TBC – $\vartheta_0 = $ const

$$F_n = \sqrt{\frac{n}{2}} \frac{\sinh\left(\sqrt{2n}\tilde{\delta}\right) + \sin\left(\sqrt{2n}\tilde{\delta}\right)}{\cosh\left(\sqrt{2n}\tilde{\delta}\right) - \cos\left(\sqrt{2n}\tilde{\delta}\right)}, \tag{B.11}$$

$$\Phi_n = \sqrt{\frac{n}{2}} \frac{\sinh\left(\sqrt{2n}\tilde{\delta}\right) - \sin\left(\sqrt{2n}\tilde{\delta}\right)}{\cosh\left(\sqrt{2n}\tilde{\delta}\right) - \cos\left(\sqrt{2n}\tilde{\delta}\right)}. \tag{B.12}$$

The asymptotical case of a semi-infinite body

$$\bar{\delta} \to \infty : F_n = \sqrt{\frac{n}{2}}, \quad \Phi_n = \sqrt{\frac{n}{2}}. \tag{B.13}$$

The asymptotical case of a negligibly thin plate

$$\bar{\delta} \to 0 : F_n \to \tilde{\delta}^{-1}, \quad \Phi_n = \frac{\sqrt{2n}}{3}\tilde{\delta}. \tag{B.14}$$

(b) TBC – q_0 = const

$$F_n = \sqrt{\frac{n}{2}} \frac{\sinh\left(\sqrt{2n}\tilde{\delta}\right) - \sin\left(\sqrt{2n}\tilde{\delta}\right)}{\cosh\left(\sqrt{2n}\tilde{\delta}\right) + \cos\left(\sqrt{2n}\tilde{\delta}\right)}, \tag{B.15}$$

$$\Phi_n = \sqrt{\frac{n}{2}} \frac{\sinh\left(\sqrt{2n}\tilde{\delta}\right) + \sin\left(\sqrt{2n}\tilde{\delta}\right)}{\cosh\left(\sqrt{2n}\tilde{\delta}\right) + \cos\left(\sqrt{2n}\tilde{\delta}\right)}. \tag{B.16}$$

The asymptotical case of a semi-infinite body

$$\bar{\delta} \to \infty : F_n = \sqrt{\frac{n}{2}}, \quad \Phi_n = \sqrt{\frac{n}{2}}. \tag{B.17}$$

The asymptotical case of a negligibly thin plate

$$\bar{\delta} \to 0 : F_n = \frac{n^2}{3}\tilde{\delta}^3, \quad \Phi_n = n\tilde{\delta}. \tag{B.18}$$

C

Infinite Chain Fractions

C.1 Fundamental Theorems of Khinchin

Suitable chain fraction. Let us consider a chain fraction, which was limited by the n-term (n-suitable chain fraction):

$$s_n = c_0 - \cfrac{1}{c_1 - \cfrac{1}{c_2 - \cdots \cfrac{1}{c_n}}} = \frac{a_n}{w_n}, \quad a_0 = c_0 = 0, \quad w_0 = 1. \quad (C.1)$$

As it is known, the classical theory of suitable chain fractions is based on three fundamental theorems of Khinchin proved by the method of mathematical induction.

Theorem 1. *The law of the formation of chain fractions looks like:*

$$\left. \begin{array}{l} a_n = c_n a_{n-1} - a_{n-2}, \\ w_n = c_n w_{n-1} - w_{n-2}, \\ n \geq 2. \end{array} \right\} \quad (C.2)$$

Also, a symbolic agreement is accepted here

$$a_{-1} = 1, \quad w_{-1} = 0. \quad (C.3)$$

Theorem 2. *The numerators and denominators of two chain fractions with the numbers $n, n-1$ are connected to each other with the following relations:*

$$w_n a_{n-1} - a_n w_{n-1} = (-1)^n, \quad n \geq 1. \quad (C.4)$$

Consequence 2.1.

$$\frac{a_{n-1}}{w_{n-1}} - \frac{a_n}{w_n} = \frac{(-1)^n}{w_n w_{n-1}}. \quad (C.5)$$

Consequence 2.2. Suitable chain fractions form a converging sequence.

Theorem 3. *The numerators and denominators of two chain fractions with the numbers $n, n-2$ are connected to each other by the following relations:*

$$w_n a_{n-2} - a_n w_{n-2} = (-1)^n c_n, \quad n \geq 1, \qquad (C.6)$$

Consequence 3.1.

$$\frac{a_{n-2}}{w_{n-2}} - \frac{a_n}{w_n} = \frac{(-1)^n c_n}{w_n w_{n-2}}. \qquad (C.7)$$

Consequence 3.2. Even and odd sequences of the suitable chain fractions are majorants (the upper limits) of the correspondent infinite chain fraction.

C.2 Generalization of the Third Theorem of Khinchin

Generalization of the proof of consequences 2.2, 3.2. Consequences 2.1, 3.1 are trivial. Consequences 2.2, 3.2 can be proved by the method of mathematical induction for the positive chain fractions:

$$a_n > 0, \quad w_n > 0. \qquad (C.8)$$

A generalization of the proof of consequences 2.2 and 3.2 for the case of an arbitrary sign of the numerator and denominator of a suitable chain fraction is given below.

Theorem 4. *Denominators of the suitable chain fractions form a monotonically growing sequence. Let us apply the method of mathematical induction to prove this statement. Let us assume that the following inequality is valid:*

$$d_n = w_n - w_{n-1} \geq 0. \qquad (C.9)$$

Let us show further (C.9) results in validity of the following inequality:

$$d_{n+1} = w_{n+1} - w_n \geq 0. \qquad (C.10)$$

One can rewrite recurrent (C.2) for the value of w_{n+1} as

$$w_{n+1} = c_n w_n - w_{n-1}. \qquad (C.11)$$

Then it follows from (C.10):

$$d_{n+1} = (c_{n+1} - 1) w_n - w_{n-1}. \qquad (C.12)$$

Let us express the value of w_{n-1} from (C.9) and substitute it into (C.12)

$$d_{n+1} = d_n + (c_{n+1} - 2) w_n. \qquad (C.13)$$

Let us write down concrete expressions for the value c_n for the spatial problem of oscillations of heat transfer:

C.2 Generalization of the Third Theorem of Khinchin

for the harmonic law
$$c_n = \frac{2}{b}(1 + f_n), \qquad (C.14)$$

for the inverse harmonic law
$$c_n = \frac{2}{b}\left(1 + \frac{\sqrt{1-b^2}}{f_n}\right). \qquad (C.15)$$

It follows from here: for the harmonic law
$$c_{n+1} - 2 = \frac{2}{b}(1 - b + f_{n+1}) > 0, \qquad (C.16)$$

for the inverse harmonic law
$$c_{n+1} = \frac{2}{b}\left(1 - b + \frac{\sqrt{1-b^2}}{f_{n+1}}\right) > 0. \qquad (C.17)$$

Let us apply the method of mathematical induction to (C.13). Let us assume that the following inequality is valid:
$$w_n > 0. \qquad (C.18)$$

Then it follows from (C.13)
$$d_{n+1} > 0, \qquad (C.19)$$

and, consequently,
$$w_{n+1} = w_n + d_{n+1} > 0. \qquad (C.20)$$

Thus, in order to finalize the proof of the inequality (C.18), it is necessary to only check up validity (for the n-suitable chain fraction) of the following two inequalities:
$$w_n > w_{n-1}, \qquad (C.21)$$
$$w_n > 0. \qquad (C.22)$$

For the value of $n = 1$, recurrent formula (C.2) yields:
$$w_1 = c_1 w_0 - w_{-1}. \qquad (C.23)$$

Above, we have also proved validity of the following equality:
$$w_{-1} = 0, \quad w_0 = 1. \qquad (C.24)$$

Further, it follows for the value of $n = 1$ from (C.14) and (C.15):
$$c_1 = \frac{2}{b}(1 + f_1) > 2, \qquad (C.25)$$
$$c_1 = \frac{2}{b}\left(1 + \frac{\sqrt{1-b^2}}{f_1}\right) > 2. \qquad (C.26)$$

From (C.14), (C.15), and (C.23), validity of such inequalities follows:

$$w_1 = c_1 > 0, \tag{C.27}$$
$$w_1 - w_0 = c_1 - 1 > 1. \tag{C.28}$$

This effectively means that the inequality

$$w_1 - w_0 > 0 \tag{C.29}$$

is also valid. Thus, validity of (C.21, C.22) was proved. Hence, the proof of Theorem 4 was also obtained. But this means in fact that consequences 2.2 and 3.2 are fair also for the chain fractions, which are included in the notation of the analytical solutions for the harmonic and inverse harmonic laws of oscillations. Therefore, we have obtained a generalization of the proof of the third theorem of Khinchin for the case of an arbitrary sign on the numerator and denominator of a chain fraction.

It should be pointed out that the proof received above is valid only for the spatial law of pulsations (where the chain fractions are real values). For the time-dependent law of oscillations (where the chain fractions are complex conjugate values), it is unfortunately impossible to obtain such a proof.

D
Proof of Divergence of the Infinite Series

It is required to prove divergence of the following infinite series:

$$S = \sum_{n=1}^{\infty} \frac{F_n}{F_n^2 + \Phi_n^2}. \tag{D.1}$$

The proof will be carried out separately for each of the limiting types of oscillations of the THTC.

D.1 Spatial Type of Oscillations

(a) TBC $\vartheta_0 = $ const: the infinite series (D.1) can be written in the following form:

$$S = \sum_{n=1}^{\infty} \frac{\tanh(n\bar{\delta})}{n}. \tag{D.2}$$

It is easy to demonstrate that the following inequality is always valid:

$$n\bar{\delta} \geq \tanh(n\bar{\delta}). \tag{D.3}$$

From here, the obvious inequality results

$$\sum_{n=1}^{\infty} \frac{n\bar{\delta}}{n} = \bar{\delta}\infty \leq S. \tag{D.4}$$

Therefore, the infinite series (D.2) is always diverging.

(b) TBC $q_0 = $ const: the infinite series (D.1) can be expressed as

$$S = \sum_{n=1}^{\infty} \frac{\coth(n\bar{\delta})}{n}. \tag{D.5}$$

From the obvious inequality
$$1 \le \coth(n\bar{\delta}), \tag{D.6}$$
the following statement directly results:
$$\sum_{n=1}^{\infty} \frac{1}{n} \le S. \tag{D.7}$$
However, the infinite series in the left-hand side of (D.7) is always diverging
$$\sum_{n=1}^{\infty} \frac{1}{n} = \infty. \tag{D.8}$$
This effectively means that the infinite series (D.5) is also always diverging.

D.2 Temporal Type of Oscillations

(a) TBC $\vartheta_0 = \text{const}$: The functions of thickness in (D.1) can be written down in the following form:
$$F_n = \sqrt{\frac{n}{2}} \frac{\sinh\left(\sqrt{2n\tilde{\delta}}\right) + \sin\left(\sqrt{2n\tilde{\delta}}\right)}{\cosh\left(\sqrt{2k\tilde{\delta}}\right) - \cos\left(\sqrt{2k\tilde{\delta}}\right)}, \tag{D.9}$$

$$\Phi_n = \sqrt{\frac{n}{2}} \frac{\sinh\left(\sqrt{2n\tilde{\delta}}\right) - \sin\left(\sqrt{2n\tilde{\delta}}\right)}{\cosh\left(\sqrt{2n\tilde{\delta}}\right) - \cos\left(\sqrt{2n\tilde{\delta}}\right)}. \tag{D.10}$$

It is easy to demonstrate validity of such inequalities:
$$F_n \ge \sqrt{\frac{n}{2}}, \tag{D.11}$$
$$F_n \le \tilde{\delta}^{-1}, \tag{D.12}$$
$$\Phi_n \le \sqrt{\frac{n}{2}}. \tag{D.13}$$

This results in the following inequality:
$$\frac{F_n}{F_n^2 + \Phi_n^2} \ge \sqrt{\frac{n}{2}} \tilde{\delta}^2. \tag{D.14}$$

However, the infinite series
$$\sum_{n=1}^{\infty} \sqrt{\frac{n}{2}} \tilde{\delta}^2 = \tilde{\delta}^2 \sum_{n=1}^{\infty} \sqrt{\frac{n}{2}} = \infty \tag{D.15}$$
is always diverging. This effectively means that in the considered particular case the infinite series (D.1) is also always diverging.

(b) TBC $q_0 = \text{const}$: The functions of thickness in (D.1) can be expressed as

$$F_n = \sqrt{\frac{n}{2}} \frac{\sinh\left(\sqrt{2n\tilde{\delta}}\right) - \sin\left(\sqrt{2n\tilde{\delta}}\right)}{\cosh\left(\sqrt{2n\tilde{\delta}}\right) + \cos\left(\sqrt{2n\tilde{\delta}}\right)}, \qquad (D.16)$$

$$\Phi_n = \sqrt{\frac{n}{2}} \frac{\sinh\left(\sqrt{2n\tilde{\delta}}\right) + \sin\left(\sqrt{2n\tilde{\delta}}\right)}{\cosh\left(\sqrt{2n\tilde{\delta}}\right) + \cos\left(\sqrt{2n\tilde{\delta}}\right)}. \qquad (D.17)$$

It is easy to show validity of the following inequalities:

$$F_n \geq \frac{n^2 \tilde{\delta}^3}{3}, \qquad (D.18)$$

$$F_n \leq \sqrt{\frac{n}{2}}, \qquad (D.19)$$

$$\Phi_n \leq \sqrt{\frac{n}{2}}. \qquad (D.20)$$

Form here, an inequality results:

$$\frac{F_n}{F_n^2 + \Phi_n^2} \geq \frac{n\tilde{\delta}^3}{3}. \qquad (D.21)$$

However, the infinite series

$$\sum_{n=1}^{\infty} \frac{n\tilde{\delta}^3}{3} = \frac{\tilde{\delta}^3}{3} \sum_{n=1}^{\infty} n = \infty \qquad (D.22)$$

is always diverging. Therefore, in the considered case the infinite series (D.1) is always diverging as well.

E
Functions of Thickness for Special Problems

At the analysis of the problem of complex heat transfer (Chap. 5), the simplified equations were used for the parameter

$$H = \frac{\hat{\vartheta}_\delta^\bullet}{\hat{\vartheta}_\delta}, \qquad (E.1)$$

that is incorporated into the parameter of the thermal effect (PTE)

$$\chi = \frac{H}{\langle h \rangle}. \qquad (E.2)$$

The correspondent corrected equations for the case of the "purely temporal" oscillations of the THTC are given below.

E.1 Heat Transfer from the Ambience

The simplified equation is

$$H = \frac{\tilde{h}_0 + \tanh\left(\tilde{\delta}\right)}{\tilde{h}_0 \tanh\left(\tilde{\delta}\right) + 1}. \qquad (E.3)$$

The corrected equation looks like

$$H^2 = \frac{\left(F_1^2 + \Phi_1^2\right)\left(\tilde{h}_0^2 + 2\tilde{h}_0 \Phi_2 + F_2^2 + \Phi_2^2\right)}{\tilde{h}_0^2 + 2\tilde{h}_0 F_2 + F_1^2 + \Phi_1^2}. \qquad (E.4)$$

Here F_1, Φ_1 are the functions of thickness determined for the TBC $\vartheta_0 = \text{const}$, F_2, Φ_2 are the functions of thickness determined for the TBC $q_0 = \text{const}$ (see Appendix B).

160   E Functions of Thickness for Special Problems

E.2 Heat Transfer from an External Semi-Infinite Body

The simplified equation looks like

$$H = \frac{K + \tanh\left(\tilde{\delta}\right)}{K \tanh\left(\tilde{\delta}\right) + 1}. \tag{E.5}$$

The corrected equation is

$$H^2 = \frac{\left(F_1^2 + \Phi_1^2\right)\left[2K^2 + 2K\left(F_2 + \Phi_2\right) + F_2^2 + \Phi_2^2\right]}{\left[2K^2 + 2K\left(F_1 + \Phi_1\right) + F_1^2 + \Phi_1^2\right]}. \tag{E.6}$$

Here

$$K = \sqrt{\frac{\lambda_w c_w \rho_w}{\lambda c \rho}}. \tag{E.7}$$

The subscript "w" relates to the semi-infinite body through which heat transfer to the main body (plate of the thickness δ) is being carried out, the rest of the notations are the same as those used in the case of heat transfer from the ambience.

Index

active period, 111–114
analytical method, 6, 69
asymptotic analysis, 57, 73, 75
averaged true heat transfer coefficient, 4, 34

Biot number, 32, 34, 35, 65, 68, 69, 85, 97, 100, 107, 108, 112, 121, 123, 125, 134
boundary problem, 29, 34, 37, 39, 78, 79, 93, 103, 121, 125

complex conjugate, 37, 40, 43, 46, 69, 73, 75
complex conjugate eigenvalues, 29, 37, 93, 103
computational algorithm, 78–80, 82–84, 89, 95, 111, 114, 117, 120
conjugate problem, 3, 11, 13, 15, 18, 20
convective–conductive heat transfer, 10, 90, 110, 134

delta-function, 51, 53, 69
differential equations, 1, 18, 20, 21
dimensionless parameters, 34, 73, 134
dropwise condensation, 122, 124–126

experimental heat transfer coefficient, 4, 121

factor of conjugation, 30, 37, 73, 95, 107, 113, 118, 122, 134
filtration property, 92

functions of thickness, 38, 103
fundamental inequalities, 34

general solution, 43, 52, 55, 77, 79, 93

harmonic function, 79, 83, 84, 89
heat conduction equation, 6, 16, 18, 19, 27, 30, 31, 34, 39, 69, 78, 79, 88, 90–93, 102, 103, 105, 125
heat transfer intensity, 32, 33, 39, 51, 52, 54, 57, 69, 75, 82, 86, 88, 89, 91, 103, 106, 108, 109, 121, 133
heat transfer processes, 1, 15, 19, 121
heat transfer surface, 2–4, 6, 10, 18, 27, 29, 30, 32, 35, 53, 69, 80, 81, 91, 93, 114, 125, 127

infinite series, 42, 50, 56, 58, 67, 77, 84
inverse harmonic function, 9, 56, 82, 84, 86, 90

Model experiment, 122

Newton's law of heat transfer, 3, 4
nucleate boiling, 12, 17–19, 85, 109, 126–131, 133–135
nucleation site density, 129–131, 133

parameter of the thermal effect, 78
periodic oscillations, 7, 10–12, 20, 54, 68, 73, 75, 121, 126, 129
phase shift, 78, 80–82
power series, 50, 76, 77, 83, 85

progressive wave, 27, 28, 30, 31, 78, 88, 90, 91, 93, 108, 117, 129, 132, 133

quadrature, 78, 79, 82–84, 89, 92

Reynolds analogy, 10, 13–15

semi-infinite body, 41, 50, 54, 97, 98, 101, 108, 112, 113, 116, 121, 134
small parameter, 42, 51, 57, 79, 81–85
smooth oscillations, 79, 86
standing wave, 90, 91
step function, 7, 35, 36, 54–56, 67–69, 111, 114, 116

Taylor series expansion, 41, 42
temperature oscillations, 2, 4, 13, 15, 16, 31, 33, 35, 37, 54, 93
thermal boundary conditions, 95
thermal resistance, 106, 108, 109
thermophysical properties, 3, 7, 10–12, 19, 20, 98, 102, 121, 126, 134
true heat transfer coefficient, 3, 29
turbulent flows, 1, 2, 10, 19

universal algorithm, 74, 75

vapor bubble, 126, 127, 129

Printing: Krips bv, Meppel
Binding: Stürtz, Würzburg